P9-BJO-569

TIMELINES OF SCIENCE
AND TECHNOLOGY
VOLUME 8

THE ATOMIC AGE

1920–1949

JOHN O.E. CLARK
WITH
MICHAEL ALLABY AND AMY-JANE BEER

www.scholastic.com/librarypublishing

Published by Grolier
an imprint of Scholastic Library Publishing,
Old Sherman Turnpike
Danbury, Connecticut 06816

Set ISBN 978-0-7172-6101-7
Volume 8 ISBN 978-0-7172-6109-3

Library of Congress Cataloging-in-Publication Data

Timelines of science and technology
p. cm.
Includes bibliographical references and index
Contents: v. 1. Origins of science — v. 2. Classical and early medieval science — v. 3. Late medieval and Renaissance science — v. 4. The Scientific Revolution — v. 5. The Industrial Revolution — v. 6. The Age of Steam — v. 7. The Age of Electricity — v. 8. The Atomic Age — v. 9. The Space Age — v. 10. The modern world.
ISBN 978-0-7172-6101-7 (set : alk paper) — ISBN 978-0-7172-6102-4 (v. 1 : alk paper) — ISBN 978-0-7172-6103-1 (v. 2 : alk paper) — ISBN 978-0-7172-6104-8 (v. 3 : alk paper) — ISBN 978-0-7172-6105-5 (v. 4 : alk paper) — ISBN 978-0-7172-6106-2 (v. 5 : alk paper) — ISBN 978-0-7172-6107-9 (v. 6 : alk paper) — ISBN 978-0-7172-6108-6 (v. 7 : alk paper) — ISBN 978-0-7172-6109-3 (v. 8 : alk paper) — ISBN 978-0-7172-6110-9 (v. 9 : alk paper) — ISBN 978-0-7172-6111-6 (v. 10 : alk paper)
1. Science-History. 2. Discoveries in science. 3. Technology-History. I. Grolier (Firm)

Q125.T587 2006
509—dc22 2005050387

For information address the publisher:
Grolier, Sherman Turnpike,
Danbury, Connecticut 06816

Printed and bound in Singapore.

FOR THE BROWN REFERENCE GROUP

Consultant: Erin Dolan, Virginia Polytechnic and State University, U.S.A.

Project Editor: Graham Bateman
Editor: Virginia Carter
Designer: Steve McCurdy
Research: Geoff Roberts
Production: Alastair Gourlay, Maggie Copeland
Editorial Director: Lindsey Lowe

PICTURE CREDITS
(t = top, b = bottom, c = center, l = left, r = right)

Cover: *Space Shuttle*, NASA Headquarters – Greatest Images of NASA; *19th-Century Steam Train*, Science Photo Library; *Galileo,* AKG-London

6t David Parker/Science Photo Library; **6b** Roger-Viollet/TopFoto; **7** David A. Hardy, Futures: 50 Years in Space/Science Photo Library; **8t** Science Museum; **9** Science Museum; **10b** Harlingue/Roger-Viollet/Topfoto; **11c** TopFoto; **12t** NMPFT; **12b** NMPFT Daily Herald Archive; **13b** Science Museum; **14b** TopFoto; **15c** NMPFT Daily Herald Archive; **15bl** TopFoto; **15br** Science Museum; **16t** Science Museum; **16b** Hulton-Deutsch Collection/Corbis; **17b** Novosti/TopFoto; **18t** Science Photo Library; **19b** TopFoto; **20t** Bettmann/Corbis; **20b** TopFoto; **21** Ron Boardman; Frank Lane Picture Agency/Corbis: **22t** NASA STI Program: **22b** Robert Hunt Library: **23t** NASA STI Program: **24t** TopFoto; **24b** Science Museum; **25** NMPFT-Walter Nurnberg; **26c** Science Museum; **26b** Robert Hunt Library; **28t** Hulton-Deutsch Collection/Corbis; **28b** Bettmann/Corbis; **29b** Bettmann/Corbis; **30b** Robert Hunt Library; **31b** Robert Hunt Library; **32t** TopFoto; **33t** TopFoto; **33b** NMPFT Daily Herald Archive; **34t** TopFoto; **35t** Bletchley Park Trust; **35b** Bettmann/Corbis; **36t** Bettmann/Corbis; **36c** Bettmann/Corbis; **37** Science Museum; **38b** Museum of Flight/Corbis; **39t** Science Museum; **40t** NMPFT Daily Herald Archive; **41** TopFoto; **42b** Bettmann/Corbis; **43b** Science Museum; **44t** Bettmann/Corbis; **44b** Science Museum; **45t** Bettmann/Corbis; **45b** Bettmann/Corbis

CONTENTS

How to Use this Set

Each of the volumes in this set covers a distinct period in our history. The periods are displayed on pages 6–7 of Volume 1, together with detailed contents. Within each volume there are three types of articles. Timeline articles list year by year the discoveries or inventions. They are arranged in horizontal bands corresponding to particular disciplines so that you can see at a glance how they relate to other areas of scientific knowledge. Each Timeline band—for example, physics or chemistry—has its own color. Interspersed among the Timeline articles are double-page Special Features, elaborating a particular topic from a Timeline. These in-depth articles focus on the background to a discovery, give information about the people involved, and explain the ways in which the discoveries or inventions have been put to use. For example, in Volume 8 (pages 28–29) "The Evolution of the Helicopter" develops a 1936 Timeline article from page 26 about helicopters. Boxed features add to the available information, often explaining scientific principles. The Key People pages at the end of each volume give full biographical details about prominent individuals or groups mentioned in the Timelines or the Features, thus weaving together in one story their achievements. Used together, the three article styles enable you to build up a comprehensive picture of the circumstances leading to a particular breakthrough moment.

Fully captioned Illustrations play a major role in this set. They include early prints, contemporary photographs, artwork reconstructions, and explanatory diagrams.

A number of cross-reference devices help you navigate through the volumes. Names of individuals with detailed biographies in the Key People pages are highlighted throughout the text using a bold typeface, for example, **Alfred Nobel**. Some individuals lived and worked in periods covered by more than one volume; and sometimes a person from an earlier or later period in history is mentioned. For this reason, a name in a bold typeface will not necessarily be found in the Key People pages of the particular volume being studied. To find out the location of a biography, consult the Set Index beginning on page 53 of each volume. Other names are printed in a bold italic typeface, for example, ***Michael Faraday***. This signifies that the person is the subject of a double-page feature article (in this case The Nature of Light, Volume 4, pages 28–29). Again, the exact page reference can be found in the Index.

At the top of most articles you will find direct cross references to special features that are relevant to a topic contained in the article.

A Glossary of some terminology used is included in each volume. It will help you if there are words that you do not fully understand. Each volume ends with a list of Further Reading and Useful Web Sites that will help you take your research further. Finally, the Set Index lists all the people and major topics covered in the complete set.

Double-page special feature articles elaborate particular topics

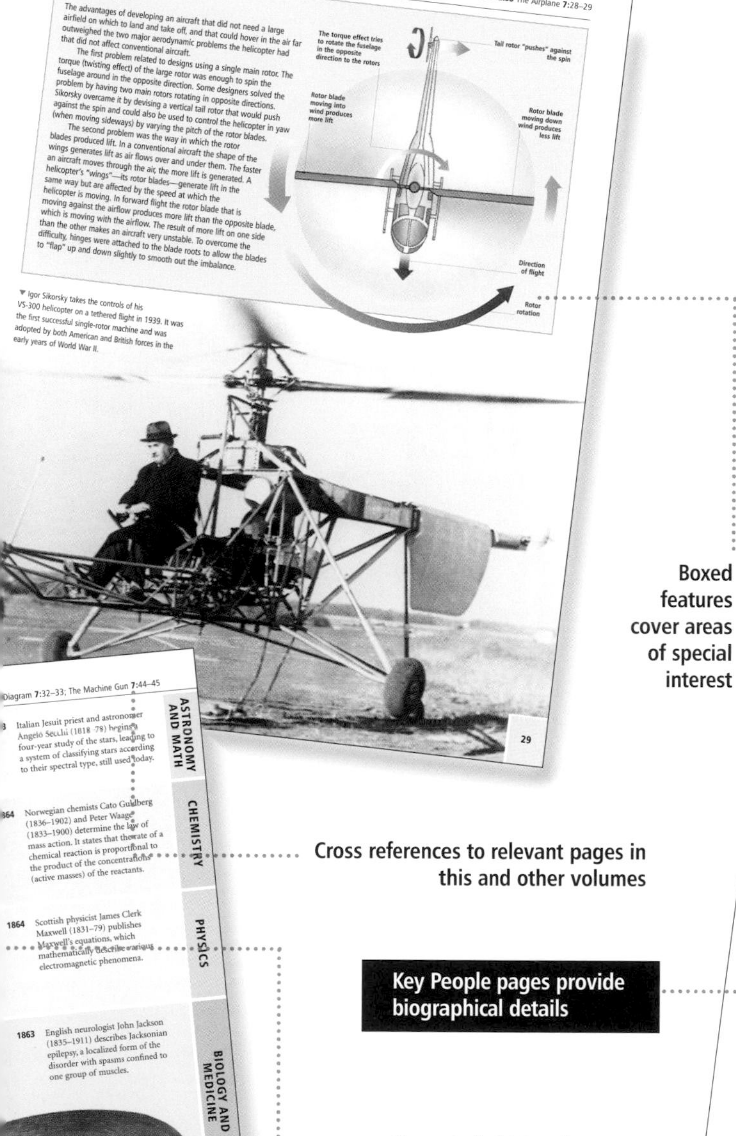

Boxed features cover areas of special interest

The Atomic Age 1920–1949

The new science of electronics continued to expand after 1920, first with the invention of television and then with the development of radar. Research into radar received a tremendous boost during World War II (1939–45) because of the need to detect enemy aircraft and to aim naval guns for long-range battles at sea. The war also concentrated efforts on rocket research, particularly in Germany, which developed its two rocket-powered "vengeance weapons" for use against targets in Britain. Alongside rapid advances in conventional aircraft, including jet engines, World War II accelerated the development of the helicopter, which was quickly adapted by the military for use in warfare. One beneficial outcome of the war was the development of methods for mass producing the antibiotic penicillin, which was needed in large quantities to treat the wounded.

But the most significant scientific development to come out of this period, and the one that probably hastened the end of the war in the Pacific, was nuclear fission. It was just three short years from the first demonstration of a controlled nuclear chain reaction in an atomic pile at Chicago University in 1942 to the detonation of the devastating atomic bombs that were dropped on Japan in 1945. Within seven years American scientists had developed the much more powerful hydrogen bomb.

The year 1947 saw the invention of the transistor, the first solid-state device, which was to revolutionize the electronics industry. It appeared just in time for the first commercial computers, which could be made faster and more powerful now that they no longer depended on unreliable vacuum tubes. Portable transistor radios and electronic calculators soon became commonplace.

Cross references to relevant pages in this and other volumes

Key People pages provide biographical details

Name in bold typeface directs reader to Key People pages

Clear captions provided for each illustration or photograph

Colored horizontal bands separate the various scientific/technological disciplines

KEY PEOPLE 1950–1979 A.D.

ASTRONOMY AND MATH

1920 American astronomer Vesto Slipher (1875–1969) detects a redshift in the light from galaxies, proving that they are receding (and providing evidence for an expanding Universe).

1920 Indian astrophysicist and nuclear physicist Meghnad Saha (1893–1956) formulates Saha's equation, which concerns the ionization of gases and helps in interpreting the spectra of stars.

CHEMISTRY

1920 Scottish chemist Arthur Lapworth (1872–1941) establishes the role played by electrons in organic chemical reactions.

1920 Belgian-born American chemist Julius Nieuwland (1878–1936) polymerizes acetylene, producing divinyl acetylene.

PHYSICS

1920 English physicist **Francis Aston** (1877–1945) formulates the so-called "whole number rule": that the mass of the oxygen isotope being defined, all the other isotopes have masses that are very nearly whole numbers.

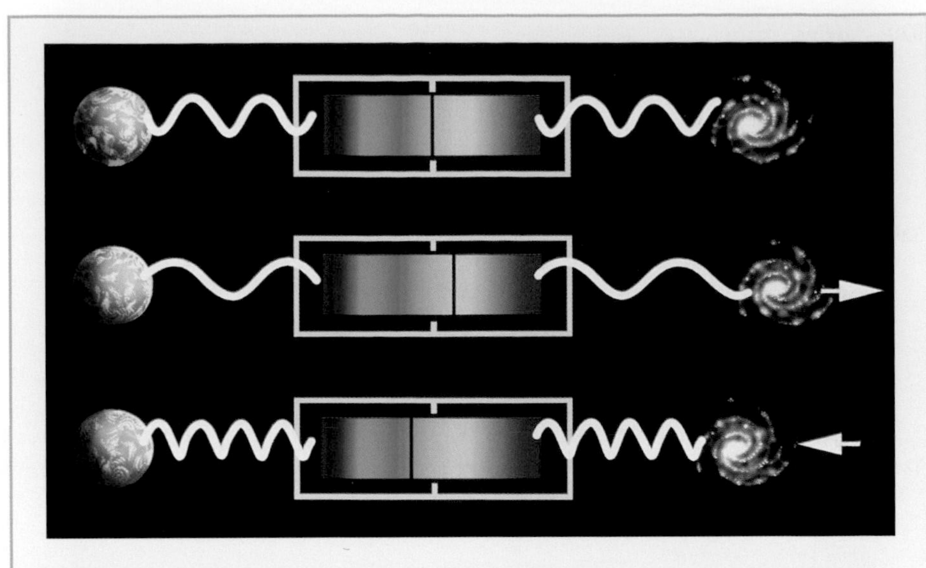

Vesto Slipher noticed in 1920 that the light coming from many galaxies was shifted toward the red end of the spectrum and concluded that they were moving away from Earth. This is an example of the Doppler effect—a shift in the apparent wavelength of radiation (such as light or sound) emitted by an object that is moving away from or toward an observer. Because they have a finite speed, the waves in front of a moving object are compressed or shortened. In the case of sound waves they produce a higher note—while the waves behind are lengthened, giving a lower note. The effect can be heard when an ambulance rushes past you sounding its siren. The same occurs with light waves, with the result that the shortened waves show light at the blue end of the spectrum, whereas the lengthened waves show as red light—the "redshift."

The upper diagram shows the spectrum of a stationary galaxy (on the right) as viewed from Earth (on the left). In the middle diagram the galaxy is moving away from Earth. This is shown by an absorption line in its spectrum that is shifted toward the red end. In the bottom diagram the galaxy is approaching, and the absorption line is shifted toward the blue end of the spectrum.

BIOLOGY AND MEDICINE

1921 Swiss psychiatrist Hermann Rorschach (1884–1922) introduces the Rorschach inkblot test for examining patients' personalities.

The Rorschach inkblot test, introduced in 1921, was intended to reveal an individual's attitudes and emotions. However, it is rarely used today.

1921 German-born American physiologist Otto Loewi (1873–1961) discovers a chemical released by stimulated nerves, which English physiologist Henry Dale (1875–1968) later identifies as acetylcholine.

1921 Canadian physiologist **Frederick Banting** (1891–1941) isolates insulin, the hormone that controls the levels of the sugar glucose in the blood.

1921 English biochemist Frederick Gowland Hopkins (1861–1947) discovers glutathione, an important requirement for cell metabolism involving oxygen (respiration).

ENGINEERING AND INVENTION

1920 The Panama Canal is officially opened by U. S. President Woodrow Wilson (1856–1924).

1920 KLM, the Dutch National Airline, flies its first scheduled service between Amsterdam in the Netherlands and London, England.

1920 American gunsmith John Thompson (1860–1940) patents and publicly demonstrates the Thompson submachine gun (Tommy gun).

1920 Station KDKA in Pittsburgh, Pennsylvania, begins the first regular radio broadcasts in the U. S.

1921 American physicist Albert Hull (1880–1966) invents the high-frequency magnetron vacuum tube, later used in microwave radar transmitters.

The Thompson submachine gun.

See also The Panama Canal **7:**40–41; The Machine Gun **7:**44–45; Subatomic Particles **8:**8–9; The Development of Radar **8:**32–33

ASTRONOMY AND MATH

1920 German-born American astronomer **Walter Baade** (1893–1960) discovers the asteroid Hidalgo, an object 18.6 miles (30-km) in diameter that follows a highly elliptical, cometlike orbit.

1920 German astronomer Max Wolf (1863–1932) observes dark clouds of interstellar matter in our galaxy, the Milky Way.

1920 English amateur astronomer and comet-hunter William Denning (1848–1931) discovers Nova Cygni.

1921 German mathematician Emmy Noether (1882–1935) begins to develop mathematical axioms ("rules") for algebra.

1921 English economist John Maynard Keynes (1883–1946) publishes his *Treatise on Probability*.

1922 Canadian astronomer John Plaskett (1865–1941) identifies the supergiant binary star known as Plaskett's star.

In 1920 Max Wolf used photography to find dark clouds of gas and dust in the Milky Way, shown here from above (top) and side on (bottom).

CHEMISTRY

1921 American chemist Thomas Midgley (1889–1944) discovers the "antiknock" properties of tetraethyl lead (it prevents preignition in gasoline engines).

1922 German chemist Hermann Staudinger (1881–1965) coins the term "macromolecule" and recognizes that substances such as rubber are natural polymers.

PHYSICS

1920 New Zealand-born English physicist **Ernest Rutherford** (1871–1937) predicts the existence of the neutron.

1921 American physicist **Arthur Compton** (1892–1962) suggests that ferromagnetism is caused by electron spin.

BIOLOGY AND MEDICINE

1921 Scottish bacteriologist **Alexander Fleming** (1881–1955) identifies the bacteria-killing enzyme lysozyme.

1922 Following experiments on dogs in 1919–20, English pharmacologist Edward Mellanby (1884–1955) establishes the use of cod-liver oil as an effective treatment for rickets in children.

1922 American biochemist Elmer McCollum (1879–1967) discovers vitamin D (calciferol).

1922 American anatomist and embryologist Herbert Evans (1882–1971) discovers vitamin E (tocopherol).

1922 Scottish physiologist John Macleod (1876–1935) and coworkers first use insulin to treat patients with diabetes.

1922 American psychologist G. Stanley Hall (1844–1924) publishes *Senescence*, one of the first books to discuss the social and psychological aspects of aging.

ENGINEERING AND INVENTION

1921 American airman John Macready (1887–1979) gives the first demonstration of spraying crops from an airplane.

1921 The first expressway in Germany (*Autobahn*), designed by Karl Fritsch, is completed.

1922 American engineer **Lee De Forest** (1873–1961) creates a sound-on-film optical recording system for movies.

1922 Australian brothers Albert and Cliff Howard produce a heavy-duty steam-powered rotary hoe.

1922 After five years of research Canadian-born American electrical engineer **Reginald Fessenden** (1866–1932) makes an improved echo sounder.

1922 The British Broadcasting Company (later Corporation)—BBC— begins regular broadcasts in Britain.

1922 American airman James Doolittle (1896–1993) makes the first coast-to-coast flight across the United States, taking 21 hours 19 minutes.

1922 Swedish inventor Nils Dalen (1869–1937) patents the Aga solid-fuel stove.

SUBATOMIC PARTICLES

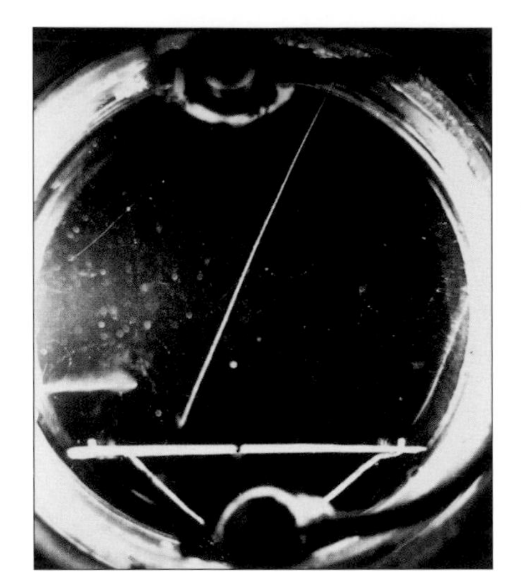

▲ A cloud chamber contains a vaporized mixture of water and alcohol that condenses when an electrically charged particle moves through it. A trail of liquid droplets is produced, marking the track of the particle. This photograph, taken in 1937, shows the track of an alpha particle (helium nucleus).

KEY DATES	
1920	Proton named
1925	Cloud chamber of 1911 developed further
1932	First antimatter particle; neutron identified
1934	Neutrino identified and named
1937	Muon discovered

By 1920 physicists knew that every atom consists of a nucleus carrying a positive electromagnetic charge surrounded by a cloud of electrons carrying negative charge. This implies that the atom is not an "elementary particle"—an object that cannot be divided into smaller constituents and that is therefore the basic building block of matter. Before long, scientists were identifying a growing list of particles much smaller than atoms.

Ernest Rutherford (1871–1937), the New Zealand-born English physicist, found that when he bombarded nitrogen atoms with alpha particles, hydrogen nuclei were released. It followed that a nitrogen atomic nucleus must contain hydrogen nuclei. In 1920 Rutherford suggested the name "proton" (from the Greek *protos* meaning "first") for the hydrogen nucleus. The mass of a proton is 1,836.12 times that of an electron, and the mass of an atom is effectively the mass of its nucleus. In the same year Rutherford suggested that the nuclei of atoms more massive than hydrogen also contain particles carrying no electromagnetic charge.

Since 1919 Rutherford had been professor of physics at Cambridge University and director of the Cavendish Laboratory. His research continued to center on smashing atomic nuclei by bombarding them with alpha particles (helium nuclei). In 1925 English physicist Patrick Blackett (1897–1974), working under Rutherford's direction, developed the cloud chamber—invented in 1911 by Scottish physicist C. T. R. Wilson (1869–1959)—into a device for recording the disintegration of atoms. But alpha particles were not powerful enough to smash large nuclei, which repelled them without disintegrating. More energetic impacts were needed, and in 1932 English physicist John Cockcroft (1897–1967) and Irish physicist Ernest Walton (1903–95) built the world's first particle accelerator at the Cavendish Laboratory. It used powerful electromagnets to accelerate protons that were then directed at a target.

During the 1920s in Berlin, German physicist Walther Bothe (1891–1957) led a team of scientists conducting experiments. They fired alpha particles at atoms of certain light elements, including beryllium, boron, and lithium. In 1930 they discovered that this bombardment resulted in the emission of highly penetrating radiation. At first, the scientists thought it was gamma radiation, although it was more penetrative than any gamma radiation they had seen.

In 1932 French physicists Irène (1897–1956) and Frédéric (1900–58) Joliot-Curie found that alpha bombardment of paraffins or similar hydrocarbons (compounds of hydrogen and carbon) resulted in the emission of protons with very high energy. Closer study of this phenomenon made it increasingly unlikely that Bothe had observed the emission of gamma radiation. English physicist James Chadwick (1891–1974), also working at the Cavendish Laboratory, finally conducted experiments proving that the emission could not have been gamma radiation. He suggested instead that the radiation

The Atom Splitter

In 1932 John Cockcroft and Ernest Walton charged a hollow metal chamber to 400,000 volts and injected protons into it. The positively charged particles were driven away from the high positive voltage along a series of tubes kept at lower voltages. They struck a piece of lithium. Alpha particles (helium nuclei) that were formed in the interaction caused flashes on an observation screen and were photographed.

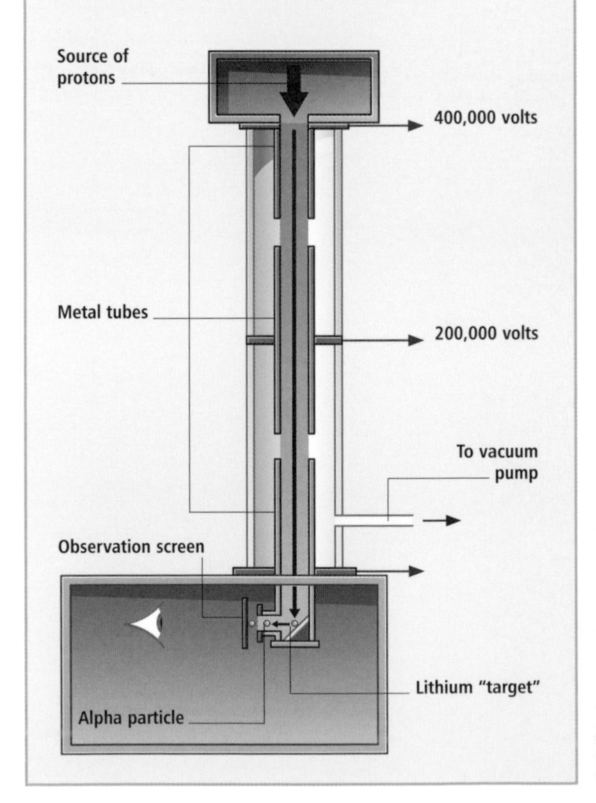

consisted of particles with approximately the same mass as the proton, but carrying no electromagnetic charge. Chadwick thought the new particle was a proton bound to an electron (a hydrogen atom); and when he bombarded boron, an element of known relative atomic mass, he was able to calculate the mass of the particle. It turned out to be 1.0087 atomic mass units, making it very slightly more massive than the proton (1.007276 a.m.u.). Because it carries no charge, the particle came to be called the "neutron." It is stable while it remains inside an atomic nucleus, but outside the nucleus a neutron decays into a proton, an electron, and an antineutrino. Protons and neutrons, the constituents of atomic nuclei, are known as "nucleons."

Wolfgang Pauli (1900–1958) was one of the greatest physicists of the 20th century. He was born in Vienna but later became a Swiss citizen. In 1930 he was studying beta radiation—a stream of electrons emitted by unstable atoms. The electrons seemed to lose energy, but no explanation could be found for the loss, which appeared to contradict one of the most fundamental laws of physics: that energy cannot be created or lost. Pauli's solution to the riddle was to propose that beta radiation also contains a previously unknown particle with the unusual properties of possessing no charge and no mass when it is at rest. Italian-American physicist **Enrico Fermi** (1901–54) confirmed the existence of such a particle in 1934 and called it the "neutrino."

Paul Dirac (1902–84), an English theoretical physicist, made important contributions to the development of quantum electrodynamics. During the late 1920s theoretical physicists were very interested in the properties of electrons. Dirac was dissatisfied with a description of the electron proposed by the German physicist Werner Heisenberg (1901–76) and came up with an alternative. The equations describing Dirac's model predicted that it should be possible for an electron to possess a positive charge. American physicist Carl Anderson (1905–91) discovered this particle in 1932, and Patrick Blackett discovered it independently in 1933. The particle was later called the "positron." It was the first antimatter particle to be identified.

In 1936, in collaboration with graduate student Seth Neddermeyer (1907–88), Anderson discovered the muon—a very unstable particle similar to an electron but 200 times more massive.

◀ A voltage multiplier, or cascade generator, installed in the Cavendish Laboratory, Cambridge, in 1937 as part of the Philips Million-Volt Accelerator. Its million-volt electric field was used to accelerate particles.

ASTRONOMY AND MATH

1923 Polish mathematician Stefan Banach (1892–1945) develops "abstract Banach spaces" from vector spaces.

1923 Austro-Hungarian rocket scientist Hermann Oberth (1894–1989) introduces the idea of escape velocity in his book *The Rocket into Interplanetary Space.*

1924 The Computing-Tabulating-Recording Company changes its name to International Business Machines (IBM).

1924 Danish astronomer **Ejnar Hertzsprung** (1873–1967) discovers DH Carinae, the first flare star (a red dwarf star that has sudden brief bright episodes).

1924 English astronomer Arthur Eddington (1882–1944) figures out the mass–luminosity relation for stars.

1925 American physicist **Robert Millikan** (1868–1953) observes cosmic rays in the upper atmosphere.

1925 Swedish astronomer Bertil Lindblad (1895–1965) proposes that our galaxy (the Milky Way) is slowly rotating.

CHEMISTRY

1923 Danish chemist Johannes Brønsted (1879–1947) and English chemist Thomas Lowry (1874–1936) independently define an acid as a substance that donates protons (hydrogen ions, H^+) to something else (a base). The base is known as a proton acceptor.

1923 Dutch-born American physicist Peter Debye (1884–1966) and German physicist Erich Hückel (1896–1980) propose their theory of electrolytes. It states that electrolytes are fully ionized in dilute solution, and that oppositely charged ions are attracted to each other.

1923 American physical chemist Gilbert Lewis (1875–1946) defines an acid as a substance that accepts electrons.

1925 English chemist Robert Robinson (1886–1975) determines the chemical structure of morphine.

PHYSICS

1923 American physicist **Arthur Compton** (1892–1962) publishes his explanation of the Compton effect, which is the increase in the wavelength (loss in energy) of an X-ray or gamma ray when it collides with electrons.

1924 English physicist Edward Appleton (1892–1965) demonstrates the existence of the ionosphere, a layer of ionized gases in the upper atmosphere, by bouncing radio waves off it.

1924 French physicist **Louis de Broglie** (1892–1987) proposes the idea of wave-particle duality: that some entities (such as electrons) sometimes behave like waves and sometimes behave as if they are particles.

BIOLOGY AND MEDICINE

1923 Austrian neurologist **Sigmund Freud** (1856–1939) introduces his concepts of the ego (the part of the mind in touch with reality) and the id (the instinctive unconscious mind).

1923 German biochemist Otto Warburg (1883–1970) demonstrates that cancerous cells absorb less oxygen than do normal cells.

1923 American physicians George (1881–1967) and Gladys Dick (1881–1963) isolate the bacteria that cause scarlet fever.

1924 German psychiatrist Hans Berger (1873–1941) detects the electrical activity of the brain and later (1929) publishes his work on electroencephalography.

1924 American pathologist George Whipple (1878–1976) and American physicians George Minot (1885–1950) and William Murphy (1892–1987) introduce the use of raw liver as a treatment for pernicious anemia.

ENGINEERING AND INVENTION

Juan de la Cierva's original autogyro.

1923 Spanish engineer Juan de la Cierva (1895–1936) successfully flies an autogyro, a helicopterlike airplane with a free-rotating horizontal propeller.

1923 Russian-born American electronics engineer **Vladimir Zworykin** (1889–1982) invents the iconoscope TV camera tube, working at the Westinghouse laboratory.

1923 The German company Benz manufactures the first diesel-engined trucks.

1923 The German company Ormig markets a spirit duplicator for making paper copies (using a dye-gelatin process).

See also The Elusive Electron **7:**16–17; Subatomic Particles **8:**8–9; The First Television **8:**12–13; The Evolution of the Helicopter **8:**28–29

Beginning in 1925, Edwin Hubble developed a classification scheme for galaxies based on their shapes. A galaxy is a huge disk of millions of stars. The so-called elliptical galaxies (three diagrams below) appear to be round, oval, or elongate depending on whether we see them flat on or side on. Spiral galaxies (upper-right sequence) reveal spiral arms as they rotate. Barred spirals (lower right) have a "bridge" of stars stretching across their centers.

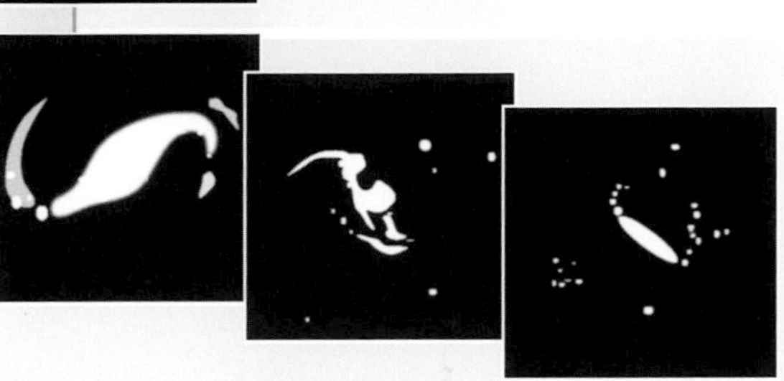

ASTRONOMY AND MATH

1925 American astronomer **Edwin Hubble** (1889–1953) introduces his classification scheme for galaxies.

CHEMISTRY

1925 German chemist Walter Noddack (1893–1960) and his wife Ida Tacke (1896–1979) discover the element rhenium by means of its X-ray spectrum.

1925 German chemist Franz Fischer (1877–1947) and Czech chemist Hans Tropsch (1889–1935) devise a process for producing oil-type fuels from coal.

PHYSICS

1925 Austrian-born Swiss physicist **Wolfgang Pauli** (1900–58) proposes his exclusion principle, which states that no two subatomic particles (such as electrons) can have the same set of quantum numbers.

1925 English physicist Patrick Blackett (1897–1974) begins experiments with colliding atoms in a cloud chamber.

1925 French physicist Pierre Auger (1899–1993) discovers the Auger effect, which is the emission of an electron from an atom with no accompanying X-ray or gamma ray.

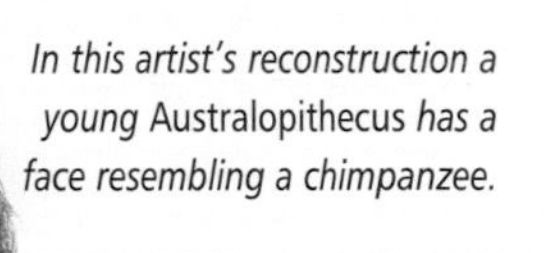

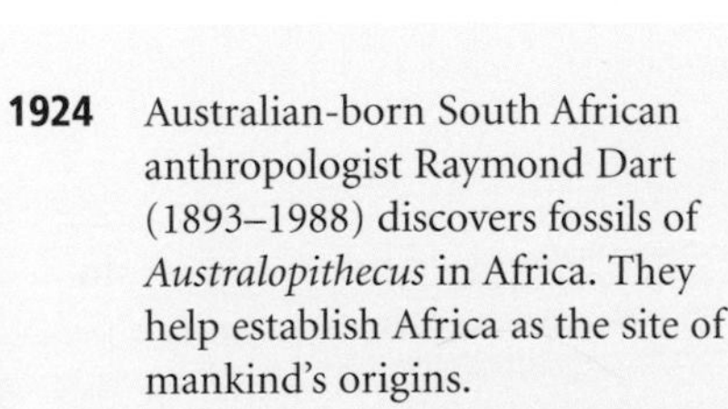

In this artist's reconstruction a young Australopithecus *has a face resembling a chimpanzee.*

BIOLOGY AND MEDICINE

1924 Australian-born South African anthropologist Raymond Dart (1893–1988) discovers fossils of *Australopithecus* in Africa. They help establish Africa as the site of mankind's origins.

1925 African-American biologist Ernest Just (1883–1941) shows that UV radiation can cause cancer.

1925 American pathologist George Whipple (1878–1976) discovers iron in red blood cells.

1925 American psychologist John Watson (1878–1958) publishes *Behaviorism,* in which he sets out for the general reader his ideas for improving society.

1925 Russian-born English biochemist David Keilin (1887–1963) names the important cellular protein cytochrome (which contains heme, a constituent of hemoglobin).

ENGINEERING AND INVENTION

1923 Scottish electrical engineer **John Logie Baird** (1888–1946) invents a television system that uses mechanical scanning.

1924 Swedish chemist Theodor Svedberg (1884–1971) develops the ultracentrifuge. It soon becomes standard equipment in biology and chemistry laboratories.

1924 American pilots Erik Nelson, Lowell Smith, Leigh Wade, and their crews make the first around-the-world flight (with refueling stops) in three U. S. Navy biplanes.

1924 Work begins on the first expressway in Italy (*autostrada*), designed by engineer Piero Puricelli (1883–1951).

1924 English watchmaker John Harwood (1893–1964) patents a self-winding wristwatch.

1924 The American company Kimberly-Clark introduces Kleenex, the world's first disposable facial tissues. Called Celluwipes, they are first marketed for the removal of cold cream.

1925 French-born American physicist Henri Chrétien (1879–1956) invents the anamorphic lens, which compresses images sideways. (Twenty-eight years later it is used for CinemaScope movies.)

THE FIRST TELEVISION

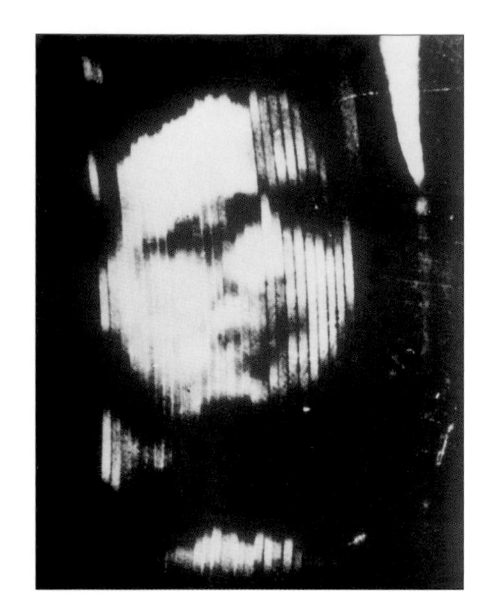

▲ This crude picture of a boy's face was made in about 1926 using John Logie Baird's mechanical television system. The flickering image was made up of only 30 vertical lines, or scans.

In October 1925 Scottish electrical engineer John Logie Baird transmitted the first television pictures in his London workshop. Unlike the electronic system later developed in the United States by Vladimir Zworykin, Baird's first camera and receiver were basically mechanical.

John Logie Baird (1888–1946) was born in the west of Scotland and educated in Glasgow. He was excused from military service at the start of World War I because of poor health. Ill health also cost him his job as an electrical engineer, and after three failed businesses he retired to live in the southern English coastal town of Hastings in 1922. It was there that he began experimenting with television.

All television cameras have some method of scanning an image. Baird used a rapidly spinning Nipkov disk patented by Polish electrical engineer Paul Nipkov (1860–1940) in 1884. It is a disk—in Baird's case made of cardboard—pierced with a spiral of holes. As the disk rotates, an observer looking through it sees an object as a series of curved lines or scans, each line produced by a different hole in the disk. The first pictures of 1925 depicted a ventriloquist's doll named Stooky Bill. The first live subject (in 1926) was an office boy from the premises below Baird's workshop, by then located in London.

At first, Baird sent his television images along wires. His "noctovisor" used infrared rays for scanning so that it captured images in the dark. By 1927 he transmitted pictures from London to Glasgow along a telephone line, and a year later he sent pictures over the Atlantic telegraph cable to New York.

In September 1929 the British Broadcasting Corporation (BBC) began experimental television broadcasts using Baird's mechanical system. The flickering images consisted of only 30 lines, later increased to 60 and eventually 240 lines. In 1932 Baird transmitted pictures by short-wave radio. The experimental broadcasts ended in 1935; and by the time commercial television broadcasting really started in Britain in 1937, the BBC had adopted the 405-line electronic system developed by the British company Marconi-EMI. But television broadcasting soon had to be suspended for the duration of World War II. Before the end of the war Baird had produced color television and three-dimensional images as well as a widescreen system (by projection) and stereophonic sound. He died before television broadcasting resumed. When it started again, it used an all-electronic system.

Scottish electrical engineer Alan Campbell-Swinton (1863–1930) figured out the principles of an electronic television system in 1908, although at that time the apparatus was not available to put his ideas

▶ Baird adjusts his early receiving apparatus. At the center of the picture is the large Nipkov disk. It has a spiral of holes that effectively scan an image as a series of lines as the disk rotates.

See also The Invention of Radio **7:**12–13; The Elusive Electron **7:**16–17

Iconoscope and Cathode-Ray Tube

The iconoscope was the first successful television camera tube, developed by Vladimir Zworykin in 1923 (patented 1938). A beam of electrons from an electron gun scans an image focused onto a photosensitive plate. Horizontal and vertical deflection plates make the electron beam scan in a series of lines from side to side. Light striking the plate gives it a positive charge, and electrons not held by the charge bounce back to a collector electrode and form the video signal. A cathode-ray tube, invented in 1897 by German physicist Ferdinand Braun (1850–1918), is the heart of a television receiver. It too has an electron gun and deflection plates and a coil to focus the electron beam onto the screen at the front of the tube. A phosphor, which gives off light when struck by electrons, coats the inside of the screen on which the scanned image builds up line by line.

ICONOSCOPE

Electron beam
Photosensitive plate
Horizontal deflection plates
Lenses
Vertical deflection plates
Image
Collector electrode
Electron gun

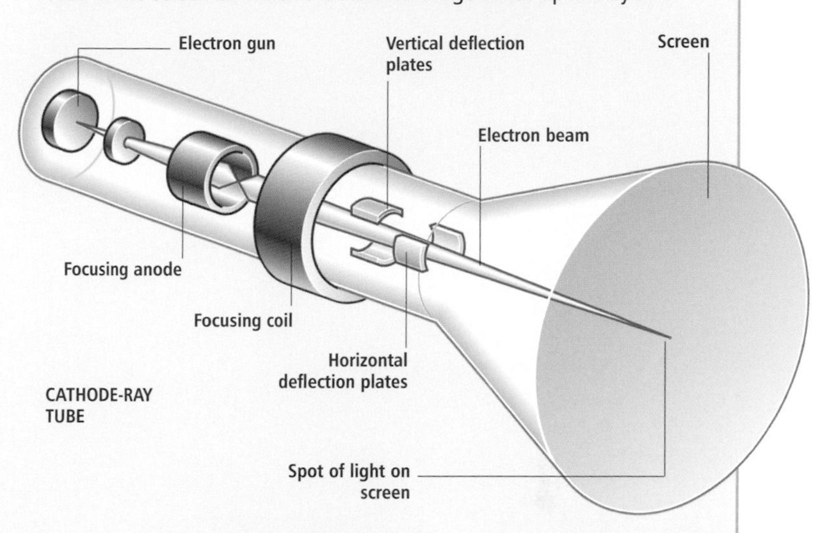

into practice. He envisaged using a cathode-ray tube in both the television camera and the receiver. He proposed that a picture signal could be transmitted along wires or, using the newly invented radio, received anywhere in range of the broadcast.

In the United States Russian-born electronics engineer **Vladimir Zworykin** (1889–1982) abandoned Baird's spinning disks and went the electronic route from the beginning. He developed his iconoscope camera tube of 1923 from a cathode-ray tube, using a beam of electrons for scanning. A lens focused light from a scene onto a so-called signal plate, which was coated with a mosaic of cesium–silver pellets. Each pellet released electrons in proportion to the amount of light falling on it, and the tube's electron beam replenished the electrons as it scanned across the signal plate. The electric current leaving the plate varied in accordance with the variations in light intensity—today we call such an output signal a video signal. American inventor Philo Farnsworth (1906–71) developed a similar camera in 1927 (patented 1930). Zworykin joined RCA and within a few years he had improved his system, but from 1939 RCA had to pay royalties to Farnsworth. In 1941 CBS made experimental color broadcasts from Station WCBW in New York, although regular transmissions in color did not begin until 1951.

▶ A British television receiver of 1936 has a vertical cathode-ray tube with a horizontal screen and an angled mirror to reflect the image toward the viewer. It cost 85 guineas (roughly equivalent to $160)—about the same price as a small family car at the time.

KEY DATES

- **1923** Zworykin invents iconoscope TV camera tube
- **1925** Baird's first television pictures
- **1929** BBC begins experimental TV broadcasts
- **1937** BBC begins commercial TV broadcasts
- **1938** Zworykin receives patent for iconoscope TV camera tube
- **1941** CBS begins experimental color TV broadcasts

ASTRONOMY AND MATH

1927 Belgian astronomer **Georges Lemaître** (1894–1966) first proposes what is known as the big bang theory of the origin of the Universe.

1927 Dutch astronomer **Jan Oort** (1900–92) confirms the rotation of the Milky Way galaxy and locates its center (in the constellation Sagittarius).

1927 German astronomers Arnold Schwassmann (1870–1964) and Arno Wachmann (1902–90) discover (using photography) the comet Schwassmann–Wachmann 1, which has a circular orbit just beyond that of Jupiter.

CHEMISTRY

1926 English chemist Christopher Ingold (1893–1970) discovers mesomerism, a phenomenon in which the structure of a chemical compound "resonates" between two alternative forms, and the substance behaves as if it were a hybrid of the two forms.

1927 English chemist William Astbury (1898–1961) takes X-ray diffraction photographs of protein fibers, the first time the technique is used to analyze noncrystalline substances.

1927 English chemist Nevil Sidgwick (1873–1952) introduces the modern theory of chemical valence, concerning the role of electrons in the various kinds of chemical bonds.

PHYSICS

1926 Austrian physicist **Erwin Schrödinger** (1887–1961) formulates the wave equation for the hydrogen atom. It is a mathematical expression that describes the behavior of an electron as it "orbits" the nucleus in a hydrogen atom.

1926 American physical chemist Gilbert Lewis (1875–1946) coins the word "photon" to describe a quantum, or "particle," of light.

1927 American physicist Clinton Davisson (1881–1958) and coworkers demonstrate the diffraction of electrons, an experiment that confirms their wavelike properties.

1927 German theoretical physicist Werner Heisenberg (1901–76) formulates his uncertainty principle.

1928 English theoretical physicist Paul Dirac (1902–84) predicts the existence of antimatter particles. (The first of them, the positron, is eventually discovered in 1932.

BIOLOGY AND MEDICINE

1926 American geneticist Hermann Müller (1890–1967) produces genetic mutations in fruit flies (*Drosophila*) by exposing them to X-rays.

1926 American biochemist James Sumner (1887–1955) is the first to crystallize an enzyme—urease.

1926 American biochemist John Abel (1857–1938) and coworkers crystallize insulin and show that it is a protein.

1927 Canadian anthropologist Davidson Black (1884–1934) identifies a new human fossil—Peking man (*Homo erectus*)—by a single tooth.

1927 English zoologist Charles Elton (1900–91) publishes *Animal Ecology* and establishes the science of ecology.

1927 Austrian-born American pathologist **Karl Landsteiner** (1868–1943) discovers the factors M and N in human blood.

ENGINEERING AND INVENTION

1926 American inventor **Robert Goddard** (1882–1945) successfully launches a liquid-fuel rocket.

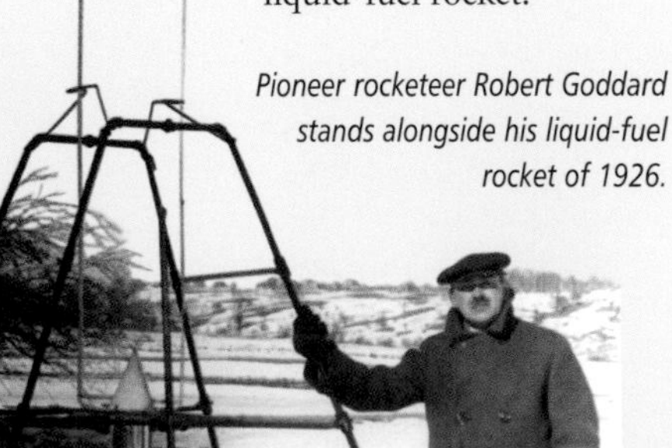

Pioneer rocketeer Robert Goddard stands alongside his liquid-fuel rocket of 1926.

1926 English aviator Alan Cobham (1894–1973) flies around the world, leaving England June 30 and returning October 1.

1926 The Delaware River Bridge (the world's longest suspension span) opens in Philadelphia, designed by Polish-born American engineer Ralph Modjeski (1861–1940). It is renamed the Benjamin Franklin Bridge in 1956.

1926 Norwegian inventor Erik Rotheim (1898–1938) invents the aerosol can.

1926 Norwegian explorer Roald Amundsen (1872–1928) and American explorer Lincoln Ellsworth (1880–1951) fly over the North Pole in the airship *Norge*, designed and piloted by Italian engineer Umberto Nobile (1885–1978).

1927 American inventor Philo Farnsworth (1906–71) makes a television camera tube (patented 1930).

1927 Canadian-born American engineer Warren Marrison (1896–1980) and Joseph Horton of Bell Telephone Laboratories invent the quartz clock.

See also Airships **7**:24–25; The Airplane **7**:28–29; Subatomic Particles **8**:8–9; Rockets **8**:16–17; Penicillin and Antibiotics **8**:20–21

ASTRONOMY AND MATH

1928 American astronomer ***Henry Russell*** (1877–1957) determines the abundance of elements in the solar atmosphere and finds that hydrogen is the principal element in the Sun's atmosphere.

1928 Austrian-born American mathematician Richard von Mises (1883–1953) publishes *Probability, Statistics, and Truth*, a philosophical approach to probability.

1928 Hungarian-born American mathematician John von Neumann (1903–57) publishes a paper outlining his minimax theorem, the basic foundation of game theory.

CHEMISTRY

1927 German-born American physicist Fritz London (1900–54) proposes a quantum theory of chemical bonding.

1928 German chemists Otto Diels (1876–1954) and Kurt Alder (1902–58) discover the Diels–Alder reaction. It is to have great value in synthetic organic chemistry for making cyclic ("ring") compounds.

1928 Russian physical chemist Nikolai Semenov (1896–1986) figures out the mechanism of chemical chain reactions. His ideas are published fully in 1934 in *Chemical Kinetics and Chain Reactions.*

PHYSICS

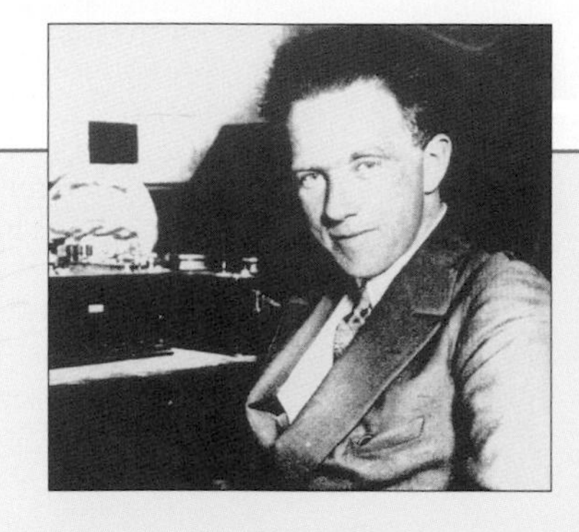

Paradoxically, Werner Heisenberg's uncertainty principle of 1927 brought more certainty to physics. The momentum of any object, including a subatomic particle, is given by multiplying its mass by its velocity. Heisenberg's uncertainty principle states that it is impossible to find exactly both the position and momentum of a particle at the same time, because the very act of taking the measurements disturbs the system and causes them to change (like trying to hit a moving target). The more exactly we determine one of them, the less exact the other one must be. Furthermore, the product of the uncertainties in the two measurements equals approximately a constant known as Planck's constant, a fundamental factor in quantum theory.

1928 Indian physicist Chandrasekhara Raman (1888–1970) describes the Raman effect: When a transparent substance is illuminated by a beam of light of a single wavelength, some of the light is scattered and has frequencies higher and lower than that of the incoming light.

BIOLOGY AND MEDICINE

1927 Portuguese neurologist António de Egas Moniz (1874–1955) detects brain tumors using cerebral angiography (in which a substance opaque to X-rays is injected into an artery serving the brain).

1928 Greek-born American physician George Papanicolaou (1883–1962) devises the Pap smear test for cancer of the cervix.

1928 Scottish bacteriologist **Alexander Fleming** (1881–1955) discovers the antibiotic penicillin.

1928 The Flying Doctor Service is introduced in Australia.

ENGINEERING AND INVENTION

1927 American electrical engineer Harold Black (1898–1983) devises the negative feedback amplifier.

1927 American aviator Charles Lindbergh (1902–74) makes the first solo flight across the Atlantic Ocean.

1928 German aviators Hermann Köhl (1881–1938) and Guenther von Hünefeld (1892–1929) and Irish army officer James Fitzmaurice (1898–1965) become the first to fly the Atlantic from east to west.

Charles Lindbergh poses in front of his plane, Spirit of St. Louis, *after his 1927 nonstop flight across the Atlantic.*

1928 American aviator Amelia Earhart (1897–1937) becomes the first woman to fly the Atlantic (not solo).

1928 American engineers Philip Drinker (1894–1972) and Louis Shaw (1886–1940) first use the iron lung, a respirator for patients with paralyzed chest muscles.

1928 American inventor Jacob Schick (1878–1937) patents the electric shaver.

Jacob Schick's electric "dry" shaver remained unchanged for many years.

Rockets

▲ Congreve rockets had a long tail with small fins for stability, like an arrow. The British army used them against Napoleon's forces and against the Americans in the Revolutionary War.

In July 1969 a huge Saturn V rocket carried three American astronauts to the Moon, where two of them landed on the surface. A little while later other rockets carried them all back to Earth. The technology that made this possible had its origins nearly eight centuries before in medieval China.

The Chinese used rockets beginning in the 1100s both for ornamental fireworks and as weapons of war. Knowledge of them quickly spread to Europe, and in 1288 the Moors used them in their attack on Valencia in Spain. Multistage rockets (with one rocket mounted on top of another) came next, and by 1715 Peter the Great had founded a rocket factory near St. Petersburg in Russia.

These early rockets were, in modern terms, solid-fuel rockets that burned gunpowder—a mixture of powdered charcoal, saltpeter, and sulfur. In 1806 the English military engineer William Congreve (1772–1828) started making rockets with an explosive warhead. When the propellant charge had burned (and the rocket had reached its target), it set off an explosive charge of gunpowder. Some Congreve rockets, launched from ramps, weighed as much as 59.5 pounds (27 kg). They reached targets 1.5 miles (2.5 km) away and were employed as artillery in the Napoleonic Wars to bombard Boulogne, France, in 1806 and Copenhagen, Denmark, in 1807. During the Revolutionary War the British built *Erebus*, a rocket-launching ship for use against American targets.

Congreve rockets had a long wooden tail, or stick, like a modern fireworks rocket. To improve accuracy and stability in flight, in about 1844 the English inventor William Hale (1797–1870) added three angled fins in the rocket exhaust, which made the missile spin in flight. Rockets no longer needed a long tail. In the mid-19th century the combatants used Hale rockets in the Mexican War and again in the American Civil War. For a while interest in military rockets declined until improvements in rocketry in the 1930s gave birth to multirocket launchers and the

KEY DATES	
1806	Congreve gunpowder rockets
1844	Hale spin-stabilized rockets
1926	Goddard's liquid-fuel rocket
1931	Gasoline/liquid oxygen rocket
1944	German V-1 and V-2 rockets

▶ In the early 1930s Hermann Oberth (left, wearing hat) developed streamlined liquid-fuel rockets that were very different in appearance from the solid-fuel "stick" rockets that preceded them.

rocket missile. Rockets did, however, find other applications. For example, in 1928 a car powered by 28 rockets (fired in sequence) reached a speed of 112 miles per hour (180 km/h) in Berlin.

Russian astrophysicist Konstantin Tsiolkovsky (1857–1935) first formalized the modern theory of rocketry in 1903. But the theory was not put into practice until 1926, when American inventor **Robert Goddard** (1882–1945) launched the first liquid-fuel rocket and ushered in a new age. The rocket, which was launched from his aunt's farm in Auburn, Massachusetts, used gasoline and liquid oxygen. It reached a speed of 65 miles per hour (105 km/h) and climbed to a height of about 41 feet (12.5 m). By 1935 Goddard's rockets could climb over 7,820 feet (2,400 m) at a speed of 620 miles per hour (1,000 km/h).

The U. S. government showed little interest in Goddard's achievements, although he continued his research. However, in Germany scientist Hermann Oberth (1894–1989) led a small research team, which by 1931 had developed a gasoline/liquid oxygen rocket. Their achievement was matched two years later by a team led by Sergei Korolev (1906–1966) in the Soviet Union. In 1930 an 18-year-old engineering student, ***Wernher von Braun*** (1912–1977), joined the German team. The team received support from the German army, and in 1936 it was allocated a new facility at Peenemünde on the shores of the Baltic Sea. There von Braun directed the research that produced the V-1 and the V-2. The V-1 was a pulsejet powered flying bomb (also known as the Doodlebug). The V-2 with its 1.1-ton (1-tonne) warhead was the first rocket-powered guided missile. The Germans used both weapons to bombard the southeast of England during the closing stages of World War II.

After the war von Braun and many other German scientists continued their work in the United States at the White Sands Proving Grounds in New Mexico. They based their initial work on captured V-2 missiles, and they launched more than 60 rockets between 1946 and 1952. A smaller rocket mounted on the nose of a V-2 produced the first two-stage rocket for high-altitude research. Progress continued rapidly in the United States and in the Soviet Union, resulting in the intercontinental ballistic missiles and the space launchers that dominated the space race for 30 years.

The Reaction Motor

Scientifically a rocket is a type of reaction motor. It makes use of ***Newton***'s third law of motion, which states that action and reaction are equal and opposite. Think of a spherical container full of expanding gases (upper diagram). The gases push equally in all directions on the walls of the container (action), and the walls resist equally (reaction), so there is no overall movement. But make a hole in one side of the container (lower diagram), and gases can escape in one direction. The equal reaction to this action pushes the vessel in the opposite direction—like a rocket.

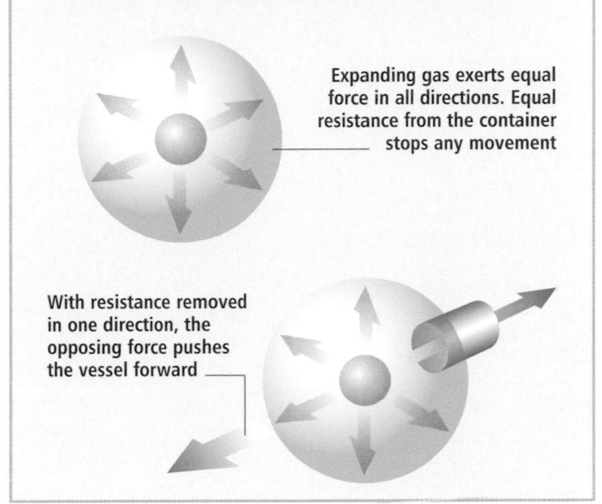

▼ The Soviet Katyusha mobile rocket launcher of World War II was an inaccurate but terrifying weapon that fired six or seven high-explosive rockets in rapid succession.

1929–1931 A.D.

ASTRONOMY AND MATH

1929 American astronomer **Edwin Hubble** (1889–1953) formulates Hubble's law, which relates the distance of a star to its velocity (of recession from Earth).

1929 Japanese geophysicist Motonori Matuyama (1884–1958) postulates that the Earth's magnetic field has undergone reversals several times in its history.

1930 American astronomer **Clyde Tombaugh** (1906–97) discovers the planet Pluto.

1930 Dutch mathematician Bartel van der Waerden (1903–96) publishes what becomes the standard work on abstract algebra, *Modern Algebra.*

Clyde Tombaugh uses the blink comparator with which he discovered the planet Pluto in 1930.

CHEMISTRY

1929 Canadian-born American physical chemist William Giauque (1895–1982) determines that natural oxygen consists of three isotopes of masses 16, 17, and 18.

1929 Irish crystallographer Kathleen Lonsdale (1903–71) uses X-ray crystallography to prove the hexagonal structure of benzene.

1930 American chemist Thomas Midgley (1889–1944) demonstrates the properties of Freon, the first CFC (chlorofluorocarbon), invented by his team in 1928 for use as a refrigerant.

PHYSICS

1929 German physicists Walther Bothe (1891–1957) and Werner Kolhorster (1887–1946) use a pair of Geiger counters to detect the direction from which cosmic rays come, and establish that cosmic rays are charged particles.

1929 German-born American physicist **Albert Einstein** (1879–1955) first announces his unified field theory, which attempts to bring all the fundamental forces into a single theory.

1930 Italian-born American physicist Bruno Rossi (1905–93) discovers the composition of cosmic rays—the primary rays being positively charged and the secondary rays being elementary particles and gamma rays.

BIOLOGY AND MEDICINE

1929 Russian-born American chemist Phoebus Levene (1869–1940) discovers the sugar deoxyribose in DNA (deoxyribonucleic acid).

1929 German biochemist Adolf Butenandt (1903–95) isolates the female sex hormone estrogen. Two years later he isolates the male hormone androsterone.

1929 Polish-born American psychiatrist Manfred Sakel (1906–57) introduces insulin shock treatment for schizophrenia.

1929 German biochemist Hans Fischer (1881–1945) determines the structure of heme (a key part of hemoglobin, the red pigment in blood) and synthesizes it.

1929 German surgeon Werner Forssmann (1904–79) performs the first successful heart catheterization (by inserting a tube into the patient's heart from a vein in the elbow).

1930 South African-born American microbiologist Max Theiler (1899–1972) develops a vaccine against yellow fever.

ENGINEERING AND INVENTION

A Wankel engine works on a four-stroke cycle. It is lighter and has fewer moving parts than the usual piston engine of similar power output. Its action is around and around. Instead of pistons it has a rotor, so there is no need for a crankshaft to convert the up-and-down motion into rotary motion.

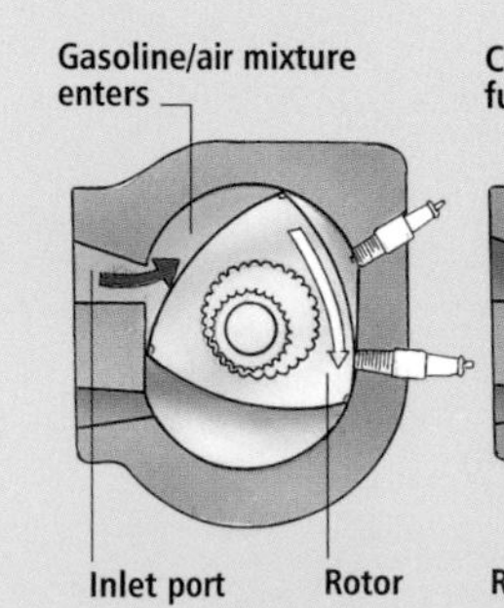

Induction
A gasoline/air mixture enters one of the chambers as the tip of the rotor goes past the inlet port.

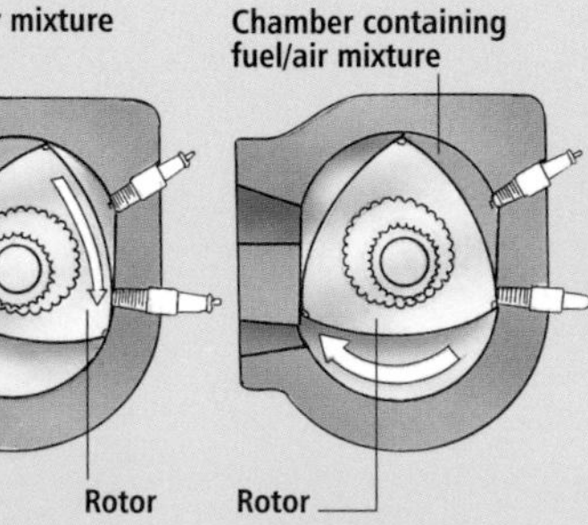

Compression
The chamber containing the mixture decreases in size as the rotor continues to turn, compressing the mixture.

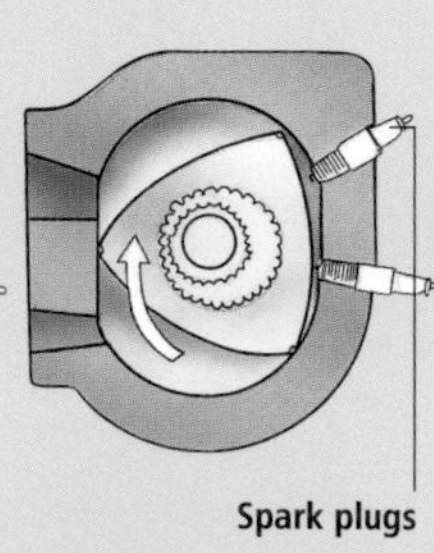

Power
A pair of spark plugs ignites the mixture, which explodes and expands, "pushing" the rotor around.

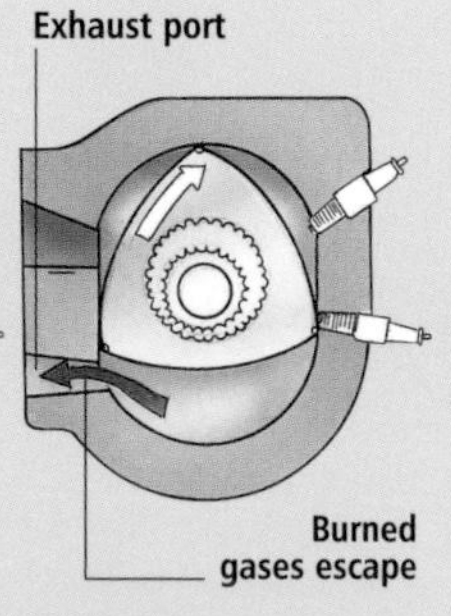

Exhaust
Burned gases escape out of the exhaust port as the leading lobe goes past.

1929 German aeronautical engineer Hugo Eckener (1868–1954) commands a Zeppelin airship that makes a 21-day around-the-world flight.

1929 German engineer Felix Wankel (1902–88) patents his rotary engine.

1929 English engineer Robert Davis (1870–1965) invents a decompression chamber for deep-sea divers and in the same year designs an escape apparatus for submarines.

1930 English engineer **Frank Whittle** (1907–96) invents the gas turbine (jet) engine.

See also The Internal Combustion Engine **6:**44–45; Artificial Fibers **8:**24–25; Nuclear Fission **8:**40–41; DNA—The Double Helix **9:**10–11

ASTRONOMY AND MATH

1930 American electrical engineer Vannevar Bush (1890–1974) builds an analog computer.

1931 Austrian-born American mathematician Richard von Mises (1883–1953) introduces the concept of sample space to probability theory.

1931 Indian-born American astrophysicist Subrahmanyan Chandrasekhar (1910–95) calculates the Chandrasekhar limit, the maximum mass a white dwarf star can possibly have (equal to about 1.4 times the mass of the Sun).

1931 Austrian-born American mathematician Kurt Gödel (1906–78) proves that any mathematical system that includes the laws of arithmetic must be inconsistent or incomplete.

1931 Italian-born American physicist Bruno Rossi (1905–93) finds that cosmic rays can travel through a piece of lead 3.2 feet (1 m) thick.

CHEMISTRY

1930 Norwegian chemist Odd Hassel (1897–1981) establishes from 1930 onward that cyclohexane consists of two stereoisomers (molecules of different atomic arrangement).

1930 American chemist Wallace Carothers (1896–1937) discovers neoprene (patented 1937). His work leads to the first commerical production of synthetic rubber and nylon.

1931 American chemist Linus Pauling (1901–94) explains that benzene is stable because its molecule has two structures that resonate (“flip” back and forth).

PHYSICS

1930 French physicist Louis Néel (1904–2000) determines the Néel temperature, the temperature above which an antiferromagnetic substance (which is not magnetic) becomes paramagnetic (slightly magnetic).

1930 Austrian-born Swiss physicist **Wolfgang Pauli** (1900–58) predicts the existence of the neutrino as a result of his studies of the production of beta rays.

1931 American radio engineer Karl Jansky (1905–50) accidentally discovers radio waves from space while investigating the nature of radio static; this will eventually lead to the science of radio astronomy.

BIOLOGY AND MEDICINE

1930 American bacteriologist Hans Zinsser (1878–1940) prepares a vaccine against typhus.

1930 Swedish biochemist Arne Tiselius (1902–71) introduces the technique of electrophoresis for separating proteins (later to become a key process in genetic fingerprinting).

1930 Swiss chemist Paul Karrer (1889–1971) determines the structure of carotene, the precursor of vitamin A, and goes on to synthesize both.

1930 American biochemist John Northrop (1891–1987) crystallizes pepsin, the protein-digesting enzyme in the stomach, and demonstrates that it too is a protein.

1931 American pathologist Ernest Goodpasture (1886–1960) devises a method of culturing viruses in chicken eggs.

1931 German chemist Adolf Windaus (1876–1959) prepares crystalline vitamin D_2 (cholecalciferol).

ENGINEERING AND INVENTION

1930 The 1,048-foot- (319.4-m-) tall Chrysler Building is completed in New York City, designed by American architect William van Alen (1883–1954).

1930 American businessman Clarence Birdseye (1886–1956) markets the first quick-frozen foods.

1930 American chemist Waldo Semon (1898–1999) invents PVC (polyvinyl chloride) plastic.

The newly built Chrysler Building towers over its neighbors in this photograph taken from the Empire State Building in 1930.

1930 English engineer Barnes Wallis (1887–1979) starts using the geodesic construction technique for aircraft, which uses light, straight structural elements.

1930 American naturalist William Beebe (1877–1962) and American engineer Otis Barton build a bathysphere submersible for deep-sea studies.

1931 American physicist Robert Van de Graaff (1901–67) invents the Van de Graaff high-voltage generator, which is later used as an atom smasher.

PENICILLIN AND ANTIBIOTICS

In the early part of the 20th century millions of people a year died from bacterial infections such as diphtheria, pneumonia, and septicemia. But the discovery of a mold that could kill germs would change all that.

▲ This is the original culture of *Penicillium* discovered by Alexander Fleming in 1928. Notice the way in which the small colonies of *Staphylococcus* do not grow close to the large *Penicillium* colony at the top.

From the late 19th century, thanks to the work of French chemist **Louis Pasteur** (1822–95) and others, scientists and medical practitioners recognized a new type of enemy—bacteria, or "germs." Many bacteria had been identified, and the conditions they caused were better understood. For example, staphylococci gave rise to boils and other skin infections, food poisoning, septicemia, and pneumonia; streptococci caused throat infections and various feverish conditions; and several bacilli produced tetanus, anthrax, diphtheria, and food poisoning.

The same kinds of organisms caused untold suffering when they infected wounds. Wound infection was the specialized area of Scottish bacteriologist **Alexander Fleming** (1881–1955). Unlike other medical researchers of the time, Fleming believed that the way to tackle infection was not to douse it with chemicals but to harness natural processes to do the job instead. So, while the German chemist Paul Ehrlich (1854–1915) devoted himself to isolating chemical therapies, Fleming was in search of biological cures. By the mid 1920s he had already made a name for himself by discovering the enzyme lysozyme (1921), produced by living cells to break down other organic material. He thought that this kind of naturally occurring substance could hold the key to fighting bacterial infection.

In 1928, after returning from vacation to his laboratory at St. Mary's Hospital in London, Fleming noticed something unusual about a culture of *Staphylococcus* that he had left developing in a petri dish. In his absence, a mold had grown in the dish, and it appeared to be killing the *Staphylococcus*. Fleming identified the strange mold as a species of *Penicillium* and discovered that the liquid it produced (penicillin) was just as effective at destroying a large number of different bacteria. More exciting still, it appeared to have no effect on healthy living tissue, so Fleming thought it could be safe to use on humans. But there were drawbacks. For a start, there were several disease-causing bacteria—notably those responsible for plague and cholera—on which it had no effect at all. (In fact, penicillin only works on gram-positive bacteria, in which the cell walls contain a thick layer of peptidoglycan—a substance that gives the bacteria shape and strength. Penicillin inhibits the formation of the peptidoglycan layer and thereby weakens the cell wall.) Even more disheartening was the fact that penicillin turned out to be very difficult to produce. For every milliliter of fluid secreted from the *Penicillium* mold, only about 0.000002 ml was active penicillin, and what tiny amounts could be extracted deteriorated very easily.

The problems seemed insurmountable at first, but in 1939 a group of scientists at Oxford University in England began following up Fleming's discovery. The team was led by Australian pathologist Howard Florey (1898–1968) and German biochemist Ernst Chain (1906–79), a refugee from Hitler's regime. By 1940 they extracted penicillin and began testing it on mice. The results were startling. A dose of penicillin enabled mice to fight off infections that otherwise would surely have killed them. Encouraged by their success, Florey began treating a human patient who was dangerously ill with staphylococcal septicemia (blood poisoning). The patient improved a little; but the doses required were too large, and Florey's laboratory was unable to produce enough to keep up

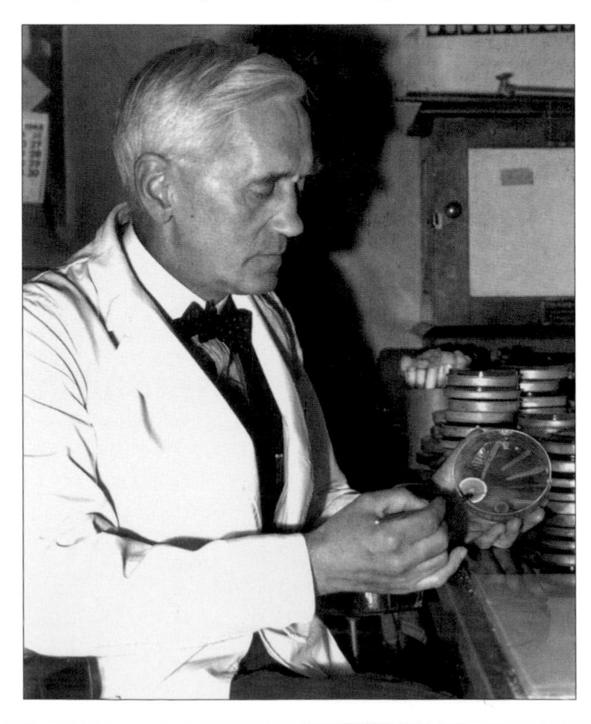

◀ Alexander Fleming served in the British Royal Army Medical Corps during World War I. His experiences influenced his subsequent search for treatments to fight wound infection, which led to his discovery of penicillin, the first antibiotic.

KEY DATES	
1877	Pasteur observes anthrax-killing bacteria
1921	Fleming discovers lysozyme in living cells
1928	Fleming identifies penicillin
1940	Mice and humans treated with penicillin
c.1943	Penicillin in mass production

the treatment. The patient died, but the results were impressive enough to convince several pharmaceutical companies of penicillin's importance. Thanks to refinements in the production technique developed by English biochemist Norman Heatley (1911–2004), the "miracle" drug was being mass-produced in the United States and Britain by about 1943.

The development could not have come at a more crucial time. World War II was raging, and wounded servicemen were the first to benefit. Thousands of lives were saved, at least on the Allied Forces' side. After the war the benefits of penicillin were spread more widely. Deaths from diseases such as anthrax, pneumonia, and tetanus and from blood infections were cut dramatically. Fleming, Florey, and Chain received the 1945 Nobel Prize for Medicine. Heatley remained an unsung hero until 1990, when his vital contribution was recognized by an honorary degree from Oxford University, where his life-saving work had taken place 50 years before.

What's in a Name?

Directly translated, the word "antibiotic" means destroyer of life. That may seem a little strange, since the main purpose of an antibiotic is to make people better. However, the life being destroyed is not that of the patient but of the bacteria causing the illness. The first antibiotic reaction was observed by **Louis Pasteur** (1822–1896) in 1877. He noticed that anthrax cultures died if they were mixed with bacteria collected from everyday sources. In 1889 French natural historian Paul Vuillemin (1861–1932) described the process by which one organism kills another as "antibiosis." From that came the modern word antibiotic, first used by Ukrainian-born American biochemist Selman Waksman (1888–1973) in 1941. He received a Nobel Prize in 1952 for his discovery of streptomycin, the first successful treatment for tuberculosis.

▲ The *Penicillium* mold (magnified here 400 times) grows in moist, nutrient-rich substances such as decaying plant matter and soil. But we are most familiar with it as the greenish-blue mold that develops on stale bread and old fruit.

ASTRONOMY AND MATH

1932 German astronomer Karl Reinmuth (1892–1979) observes the first Earth-crossing Apollo asteroid. It is lost soon after and not rediscovered until the 1970s.

1932 American astronomer Walter Adams (1876–1956) reports that the atmosphere of Venus is composed mainly of carbon dioxide.

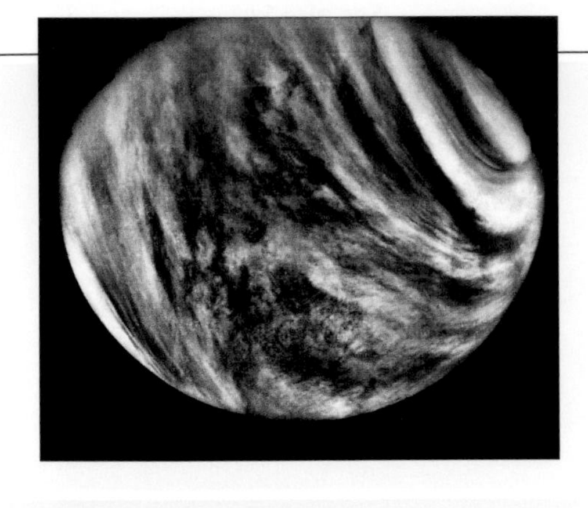

In 1974 the Mariner 10 *space probe confirmed the 1932 report of the presence of carbon dioxide in the atmosphere of Venus, as predicted by Walter Adams. This picture has been color enhanced to simulate Venus's natural color.*

1933 Swiss astrophysicist Fritz Zwicky (1898–1974) proposes that space must contain invisible "dark matter" (to account for the total mass in the Universe).

CHEMISTRY

1932 American chemist **Harold Urey** (1893–1981) and coworkers isolate deuterium (heavy hydrogen, the hydrogen isotope of mass 2). Two years later Australian physicist Mark Oliphant (1901–2000) discovers tritium (the hydrogen isotope of mass 3).

1932 German chemist Adolf Windaus (1876–1959) determines the structure of the steroid cholesterol.

1932 Synthetic rubber is first marketed in the United States (by the DuPont Company) under the name Duprene, later renamed Neoprene.

1933 English chemist Charles Coulson (1910–74) and German physicist Erich Hückel (1896–1980) independently propose the molecular orbital theory to describe the behavior of bonding electrons in chemical compounds.

PHYSICS

1932 English physicist James Chadwick (1891–1974) discovers the neutron. Together with the electron and the proton it completes the set of major subatomic particles.

1932 American physicist Carl Anderson (1905–91) announces the existence of the positron (positive electron). It is the first example of a particle of antimatter.

1932 English physicist John Cockcroft (1897–1967) and Irish physicist Ernest Walton (1903–95) carry out the first nuclear fission by bombarding lithium with protons, using a linear accelerator they invented.

BIOLOGY AND MEDICINE

1932 American physiologist Walter Cannon (1871–1945) introduces the concept of homeostasis, in which the body's systems act together to maintain a condition of balance.

1932 Swedish biochemist Axel Theorell (1903–82) isolates crystalline myoglobin, the oxygen-carrying protein in muscle tissue.

1932 Anthropologist George Lewis discovers fossils of the hominoid *Ramapithecus* in northern India.

1932 Hungarian-born American biochemist Albert Szent-Györgyi (1893–1986) and American biochemist Charles King (1896–1986) independently prove hexonuric acid is vitamin C; it is isolated and renamed ascorbic acid.

1932 German bacteriologist Gerhard Domagk (1895–1964) discovers the first sulfa drug, sulfanilamide (trade name Prontosil). It is first used in 1935 against a streptococcal infection.

1932 American physician Armand Quick (1894–1978) develops the Quick test for blood clotting, used when administering anticoagulants.

ENGINEERING AND INVENTION

The Sydney Harbor Bridge.

1932 The Sydney Harbor Bridge, designed by English engineer Ralph Freeman (1880–1950), is completed. At the time it has the longest single span of any bridge in the world.

1932 American physicist Ernest Lawrence (1901–58) successfully operates the first cyclotron, one of the first particle accelerators, or atom smashers.

1932 Dutch physicist Frits Zernike (1888–1966) invents the phase-contrast microscope.

1932 American electrical engineer William Kouwenhoven (1886–1975) has the idea for a defibrillator to restart the heart of a patient with a heart attack.

ASTRONOMY AND MATH

1933 American astronomer Donald Menzel (1901–76) determines the existence of oxygen in the Sun's corona.

1934 Swiss astrophysicist Fritz Zwicky (1898–1974) and German-born American astronomer **Walter Baade** (1893–1960) predict the existence of neutron stars from their analysis of supernovas.

A tiny neutron star lies at the very center of the Crab nebula, the remnant of a supernova explosion that was observed by Chinese astronomers in 1054. Fritz Zwicky and Walter Baade predicted the existence of such stars in 1934. They are the densest and smallest stars, consisting of tightly packed neutrons (so-called degeneracy matter). They have a mass similar to that of the Sun but are only about 12 miles (20 km) across. Rapidly spinning neutron stars become pulsars, emitting high-energy pulses of radio waves. This image was taken in 1999 by the orbiting Chandra X-Ray Observatory (CXO).

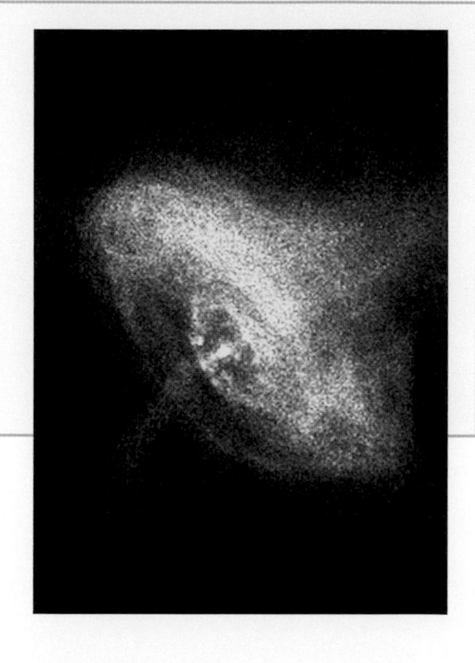

CHEMISTRY

1933 Polish chemist Tadeus Reichstein (1897–1996) and, working together, English chemists Norman Haworth (1883–1950) and Edmund Hirst (1898–1975) synthesize vitamin C (ascorbic acid) a year after it was first isolated. It is the first vitamin to be artificially produced.

1934 French physicists Irène (1897–1956) and Frédéric Joliot-Curie (1900–58) produce the first artificial radioisotope by bombarding aluminum with alpha particles.

PHYSICS

1934 English physicist Geoffrey Taylor (1886–1975) explains that crystal dislocations cause crystalline materials to fail (break) under lower than expected stresses.

1934 Dutch physicist Hendrik Casimir (1909–2000) puts forward a theory to account for the phenomenon of superconductivity, in which some materials have zero electrical resistance at low temperatures.

1934 Russian physicist Pavel Cherenkov (1904–90) discovers Cherenkov radiation, a blue light emitted when charged particles pass through water at a speed greater than the speed of light traveling in water.

BIOLOGY AND MEDICINE

1933 German chemist Richard Kuhn (1900–67) and coworkers isolate vitamin B_2 (riboflavin). The same year American chemist Roger Williams (1893–1988) discovers another B complex vitamin, pantothenic acid.

1933 The Tasmanian wolf (or thylacine) becomes extinct in the wild.

1933 American biochemist George Wald (1906–97) discovers vitamin A in the retina of the eye, later shown to be a precursor of rhodopsin (a red photosensitive pigment).

1934 American scientist Royal Raymond Rife (1888–1971) tests a cancer-curing treatment using radio waves on human patients.

1934 German biochemist Adolf Butenandt (1903–95) identifies the female sex hormone progesterone.

1934 Danish biochemist Henrik Dam (1895–1976) discovers a factor needed for blood clotting, which a year later he calls vitamin K. American chemist Robert Williams (1886–1965) isolates vitamin B_1.

ENGINEERING AND INVENTION

1933 English engineer Alan Blumlein (1903–42) patents a system of stereophonic sound recording.

1933 The first modern airliner, the Boeing 247, enters service.

1933 English aviator Alan Cobham (1894–1973) devises a system for inflight refueling of airplanes.

1934 The French company Citroën launches the first mass-produced front-wheel-drive car.

1934 American engineer Laurens Hammond (1895–1973) designs the electronic organ that bears his name.

1934 American firearms engineer John Garand (1888–1974) designs the semiautomatic Garand M1 rifle, soon to become standard issue in the U. S. Army.

1934 American psychologist B. F. Skinner (1904–90) invents the Skinner box for studying the psychology of animals.

1934 The *Queen Mary*, the world's largest ocean liner, is launched in Glasgow, Scotland. The liner's maiden voyage takes place in 1936.

1934 English inventor Percy Shaw (1890–1976) patents "cat's eyes" (reflecting road markers).

ARTIFICIAL FIBERS

▲ Wallace Carothers first developed nylon in 1935 while working for DuPont. He directed the company's research program, which also produced the synthetic rubber neoprene.

For centuries weavers had only four fibers for making cloth: silk, wool, cotton, and flax (for linen). They also used jute and hemp to produce coarse cord and sacking. The first attempts at creating an artificial fiber tried to copy silk, the most expensive of the natural ones.

Natural silk consists of the plant product cellulose. Methods of copying it involve dissolving cellulose (in the form of wood pulp or cotton) to make a solution that is forced through a tiny hole to make a strand. Chemicals harden the strand to form a fiber. The first patent for such a process—in the early 1800s—went to a Swiss chemist named Georges Audemars, who probably coined the term "artificial silk." In 1883, while searching for a material to make the filaments for his newly invented electric lamp, English physicist Joseph Swan (1828–1914) patented a process for making cellulose fibers by dissolving nitrocellulose (guncotton) in acetic acid (ethanoic acid) and forcing it through many small holes.

In France in 1884 another chemist, Hilaire de Chardonnet (1839–1924), developed a similar product. He was studying diseases in silkworms when he decided to try to imitate the silk-making process. He made cotton waste into cellulose acetate (ethanoate) and dissolved it in a solvent. He forced the viscous (sticky) solution through a mesh of small holes called spinnerets (named for the similar structures on the abdomens of silkworms and spiders). Originally known as Chardonnet silk, the fiber was later called acetate rayon.

Chemists working in Germany produced a patented product called *Glanzstoff* by dissolving cellulose in a mixture of copper sulfate and ammonium hydroxide—the so-called cuprammonium process. In 1892 in England chemists Edward Bevan (1856–1921) and Charles Cross (1855–1935) dissolved cellulose in a mixture of sodium hydroxide and carbon disulfide. Then they forced the solution through spinnerets and regenerated the cellulose as fiber by running the strands through a bath of sulfuric acid. This became known as the viscose process, and the product was therefore viscose rayon.

KEY DATES	
1883	Swan's process for artificial silk
1884	Chardonnet's acetate rayon
1892	Bevan and Cross make viscose rayon
1935	Nylon 66
1941	Dacron (Terylene)

The first totally artificial fiber was produced in 1935 by American chemist Wallace Carothers (1896–1937). It was a polymer (in fact, a polyamide). Carothers named it nylon 66 because the two starting chemicals—the monomers—each contained six carbon atoms. His company, DuPont, released the then secret product in 1938, a year after Carothers had committed suicide following years of struggling with depression. DuPont and other companies have produced several other kinds of nylon since then.

In 1941 chemists John Whinfield (1901–66) and James Dickson made a different type of polymer, a polyester with the trade names Dacron and Terylene. It is formed from terephthalic acid and ethylene glycol, and is often blended with a natural fiber such as wool. More hardwearing than rayon and more resistant to heat than nylon, it also holds color better than either of these products. In the 1950s American chemists produced

▶ In the 1930s the British company Courtaulds advertised their artificial silk as "The Finest Examples of Rayon Today." It was used mainly for making women's underwear.

Orlon, made from a cyanide compound and used for wool-type fabrics and artificial fur. Acrilan, another kind of acrylic fiber, dates from about the same time.

Swan had developed cellulose fibers for use as carbon filaments in electric lamps. In 1964 chemists at the American company Hercules and the British company Courtaulds independently reinvented carbon fibers. Manufacturers mix them with various plastics to make composite materials of great strength and versatility. Carbon-fiber composites are made into leisure products such as golf clubs, tennis rackets, and sailboat masts, as well as industrial items such as turbine blades and helicopter rotors.

Other inorganic substances used to make fibers include glass and asbestos. Glass fibers can be woven into cloth or used to reinforce carpets and materials for tents. They can also be added to synthetic resins to make the composite material fiberglass, which finds many applications, including lightweight boat hulls and car bodies. Cloth woven from asbestos fibers is fireproof and has been used for making gloves and protective clothing for firefighters.

▲ Invented in 1941, Dacron (Terylene) did not go into full production until the 1950s. Fabrics made from it are very hardwearing and keep their shape after washing. It is often blended with natural fibers such as wool.

ASTRONOMY AND MATH

1935 American radio engineer John Dellinger (1886–1962) confirms that sunspot activity causes interference with radio signals on Earth.

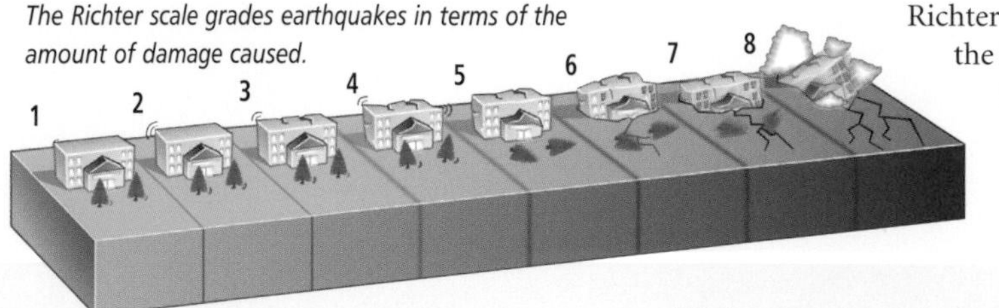

The Richter scale grades earthquakes in terms of the amount of damage caused.

1935 American oceanographer William Ewing (1906–74) starts a seismic (earthquake) study of the sea bottom.

1935 American seismologist Charles Richter (1900–85) devises the Richter scale to measure earthquake intensity.

1935 German-born American astronomer Rupert Wildt (1905–76) detects that the large planets have ammonia and methane in their atmospheres.

1936 American mathematician Alonzo Church (1903–95) demonstrates that there is no single way in which a mathematical statement can be verified or proven.

CHEMISTRY

1935 Swiss chemist Paul Karrer (1889–1971) determines the structure of vitamin B_2 (riboflavin) and synthesizes it.

1935 American biochemist William Rose (1887–1985) discovers threonine, the last of the so-called essential amino acids.

Some simple polythene items dating from 1935 are shown alongside the chemist's laboratory notebooks.

PHYSICS

1935 German-born American physicists Fritz (1900–54) and Heinz (1907–70) London announce the so-called London equations concerning superconductors.

1935 Canadian-born American physicist Arthur Dempster (1886–1950) discovers the fissile isotope uranium-235. It is used later in atom bombs.

BIOLOGY AND MEDICINE

1935 Austrian zoologist **Konrad Lorenz** (1903–89) discovers imprinting in animals (in which young learn to follow a parent using visual and auditory stimuli) and founds the science of ethology (animal behavior).

1935 American biochemist Wendell Stanley (1904–71) crystallizes tobacco mosaic virus.

1935 German-born American biochemist Rudolf Schoenheimer (1898–1941) uses deuterium (heavy hydrogen) as a tracer to follow biochemical reactions.

1935 Portuguese neurologist António de Egas Moniz (1874–1955) introduces prefrontal lobotomy as a treatment for certain personality disorders.

1936 The last-known Tasmanian wolf, or thylacine (*Thylacinus cyanocephalus*), dies at Hobart Zoo in Tasmania, Australia.

1936 The first giant panda (*Ailuropoda melanoleuca*) is captured alive in the wild by clothing designer Ruth Harkness, widow of adventurer William Harkness.

ENGINEERING AND INVENTION

1935 German engineer **Willy Messerschmitt** (1898–1978) designs the Bf-109 single-seat fighter airplane.

1935 Scottish physicist Robert Watson-Watt (1892–1973) invents radar, which is also being developed in Germany.

1935 The coin-operated parking meter, invented by American editor Carlton Magee (1873–1946), is first introduced in the U. S.

1935 American businessman Charles Darrow (1889–1967) patents the board game Monopoly.

1935 American musicians and amateur photographers Leopold Godowsky, Jr. (1900–83) and Leopold Mannes (1899–1964) invent Kodachrome transparency film.

1936 German engineer Heinrich Focke (1890–1979) builds and successfully demonstrates a two-rotor helicopter.

1936 The Boulder Dam is completed on the Colorado River. Behind it Lake Mead becomes the world's largest reservoir.

The concrete of the Boulder Dam shines under floodlights at night.

See also The Development of Radar **8:**32–33; Nuclear Fission **8:**40–41; The First Computers **8:**44–45; Radio Telescopes **9:**22–23

Astronomy and Math

1936 English mathematician and logician **Alan Turing** (1912–54) "invents" the hypothetical Turing machine, which determines whether or not a problem can be solved using a computer. A year later he devises a mathematical theory of computing and publishes his paper "On Computable Numbers."

1936 Danish seismologist Inge Lehmann (1888–1993) proposes that the Earth's inner core is solid and surrounded by liquid metal.

1937 American pioneer radio astronomer Grote Reber (1911–2002) builds the first steerable radio telescope in his backyard.

1937 Russian mathematician Ivan Vinogradov (1891–1983) proves that every large even number can be expressed as the sum of four prime numbers.

1937 American electrical engineer George Stibitz (1904–95) of Bell Laboratories makes a binary adding machine (the Model K).

Chemistry

1935 British chemists at ICI (Imperial Chemical Industries) develop a reproducible high-pressure synthesis of polyethene (also known as polyethylene or polythene).

1936 American chemist Robert Williams (1886–1965) announces the structure of vitamin B_1 (thiamin) and its synthesis.

1937 Italian-born American physicist **Emilio Segrè** (1905–89) and Italian mineralogist Carlo Perrier (1886–1948) find technetium in a sample of molybdenum that had been bombarded by deuterons in the cyclotron at Berkeley, California.

Physics

1936 Danish physicist **Niels Bohr** (1885–1962) proposes a liquid-drop model of the nucleus.

1936 American physicists **George Gamow** (1904–68) and Edward Teller (1908–2003) put forward a theory to account for beta decay by a radioactive isotope.

1936 American physicist Carl Anderson (1905–91) discovers the muon (mu-meson) in cosmic rays. The muon is a negatively charged particle some 200 times as massive as the electron.

Biology and Medicine

1937 The kouprey *(Bos sauveli)*, a wild ox, is discovered in Cambodia.

1937 German-born English biochemist Hans Krebs (1900–81) explains the citric acid cycle, also called the Krebs cycle. It is a series of reactions that are fundamental to the metabolism in aerobic (oxygen-consuming) organisms.

1937 American chemist Michael Sveda (1912–99) accidentally discovers the artificial-sweetening properties of cyclamates.

1937 American biochemist Conrad Elvehjem (1901–62) isolates the B complex vitamin nicotinic acid (niacin).

1937 Swiss-born Italian pharmacologist Daniel Bovet (1907–92) identifies the first antihistamine substance that is effective in treating allergies.

1937 Italian physician Ugo Cerletti (1877–1963) introduces ECT—electroconvulsive, or "shock," therapy—to treat patients with schizophrenia.

Engineering and Invention

1936 The Supermarine Spitfire fighter airplane, designed by English engineer Reginald Mitchell (1895–1937), makes its maiden flight.

1937 The world's first pressurized aircraft, the Lockheed XC-35, makes its maiden flight.

1937 San Francisco's Golden Gate Bridge, designed by American engineer Joseph Strauss (1870–1938), is completed.

1937 The Swiss company Nestlé begins marketing instant coffee (Nescafé).

In this early type of photocopier using the xerography process, the item to be copied is moved across the light source (instead of the light moving to scan the item as in most modern machines). A lens focuses an image of the item onto a drum (1), where it produces an electrostatic copy as a pattern of electric charges. Particles of charged toner powder stick to the pattern (2) and transfer to a sheet of blank paper as the drum rotates (3). A heater partly melts the toner to "fix" the image onto the paper (4).

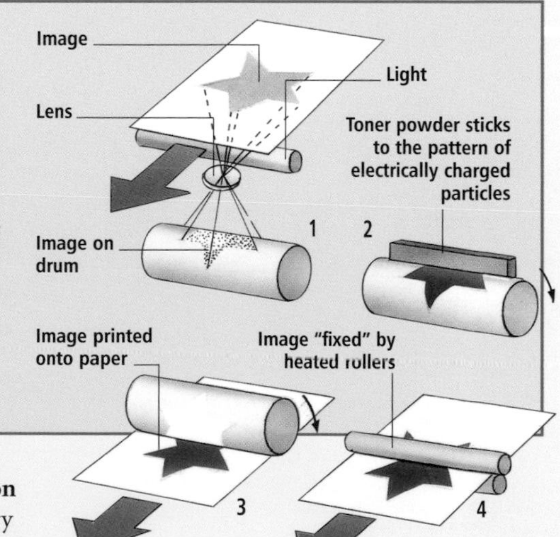

1937 American physicist **Chester Carlson** (1906–68) invents xerography, a dry photocopying process (patented 1940).

THE EVOLUTION OF THE HELICOPTER

▲ Argentinian inventor Raul Pateras Pescara was one of the early helicopter pioneers. The many-bladed rotors on his third helicopter of 1924 hauled it into the air for a record-breaking ten minutes. It is seen here on a test flight in France.

As early as 1480 Italian artist and inventor Leonardo da Vinci (1452–1519) sketched a design for a craft that carried a large, vertically mounted screw. He reasoned that if the screw rotated fast enough, the craft would rise into the air. Unfortunately, he did not know about the phenomenon of torque, which would make the craft rotate while the screw remained still.

KEY DATES	
1877	Model steam-powered machine
1907	Experimental gasoline-engined helicopter
1939	First practical single-rotor machine (by Sikorsky)
1942	U. S. Army takes delivery of Sikorsky helicopters

Even if Leonardo had understood torque, there were no motors available in his lifetime to spin the screw. Four centuries later, in 1877, Italian physician Carlo Forlanini (1847–1918) and Frenchman Gustave de Ponton d'Amécourt tried to make models of steam-powered helicopters. Forlanini's model had a pair of rotors turning in opposite directions on the same shaft. Driven by a small steam engine, they lifted the craft to a height of 49 feet (15 m), where it hovered for nearly a minute

By the beginning of the 20th century the gasoline engine was available. In 1905 English engineer E. R. Mumford patented a machine with six 25-foot (7.5-m) propellers and a bamboo airframe. In 1912 the machine, tethered by ropes, lifted a pilot 10 feet (3 m) into the air. French engineer Louis Bréguet (1880–1955) and his brother Jacques also tried to use a gasoline engine in a helicopter. In 1907 they built a machine with four rotors. Each blade resembled pairs of biplane wings, giving 32 rotor "blades" in all. With one of their assistants on board, the unwieldy Bréguet craft (also tethered) rose to a height of 24 inches (60 cm) for about a minute. Two months later the first free flight in a helicopter was made by French bicycle engineer Paul Cornu (1881–1944) at Lisieux in northwest France. His machine had two rotors, but its longest flight lasted just 20 seconds, and it only lifted 6.5 feet (2 m) off the ground.

Neither of the French machines successfully tackled the problem of directional stability—crucial for sustained flight. Between 1908 and 1912 Russian-born American engineer Igor Sikorsky (1889–1972) set his mind to solving the problem. He emigrated to the United States in 1919, where he continued to build experimental machines. The challenge was also taken up by others: American electrical engineer Peter Cooper Hewitt (1861–1921) in 1918; Louis Bréguet and René Dorand (1898–1981) in France in 1935; and Heinrich Focke (1890–1979) in Germany in 1936. Focke's Fa-61 twin-rotor machine could fly backward as well as forward at a speed of 75 miles per hour (120 km/h) and a height of over 7,800 feet (2,400 m). It set an endurance record of 1 hour 20 minutes. In 1938 German pilot Hanna Reitsch (1912–79) flew the machine inside the Deutschlandhalle in Berlin.

Finally, in 1939 Sikorsky flew the first practical single-rotor machine. It could take off vertically and reached a forward speed of 43 miles per hour (70 km/h). It had an enclosed cabin and a latticework tail carrying a small vertical propeller. This overcame the longstanding problem of torque in a single-rotor machine, since it kept the whole machine from rotating. During the early years of World War II American and British forces used improved versions of the Sikorsky VS-300 helicopter: The U. S. Army acquired its first machines in May 1942, and the British Royal Navy took delivery of the more powerful Sikorsky R-4 in 1943. The military helicopter came into its own during the Korean War (1950–53), transporting troops and casualties. It took on a new role as aerial artillery with the advent of the awesome helicopter gunships used by American forces in the Vietnam War (1954–75).

◀ Heinrich Focke's helicopter of the late 1930s had two three-bladed rotors and a conventional propeller that gave it a forward speed of 75 miles per hour (120 km/h). It captured the world endurance record with a time in the air of 1 hour 20 minutes.

The Helicopter

The advantages of developing an aircraft that did not need a large airfield on which to land and take off, and that could hover in the air far outweighed the two major aerodynamic problems the helicopter had that did not affect conventional aircraft.

The first problem related to designs using a single main rotor. The torque (twisting effect) of the large rotor was enough to spin the fuselage around in the opposite direction. Some designers solved the problem by having two main rotors rotating in opposite directions. Sikorsky overcame it by devising a vertical tail rotor that would push against the spin and could also be used to control the helicopter in yaw (when moving sideways) by varying the pitch of the rotor blades.

The second problem was the way in which the rotor blades produced lift. In a conventional aircraft the shape of the wings generates lift as air flows over and under them. The faster an aircraft moves through the air, the more lift is generated. A helicopter's "wings"—its rotor blades—generate lift in the same way but are affected by the speed at which the helicopter is moving. In forward flight the rotor blade that is moving against the airflow produces more lift than the opposite blade, which is moving with the airflow. The result of more lift on one side than the other makes an aircraft very unstable. To overcome the difficulty, hinges were attached to the blade roots to allow the blades to "flap" up and down slightly to smooth out the imbalance.

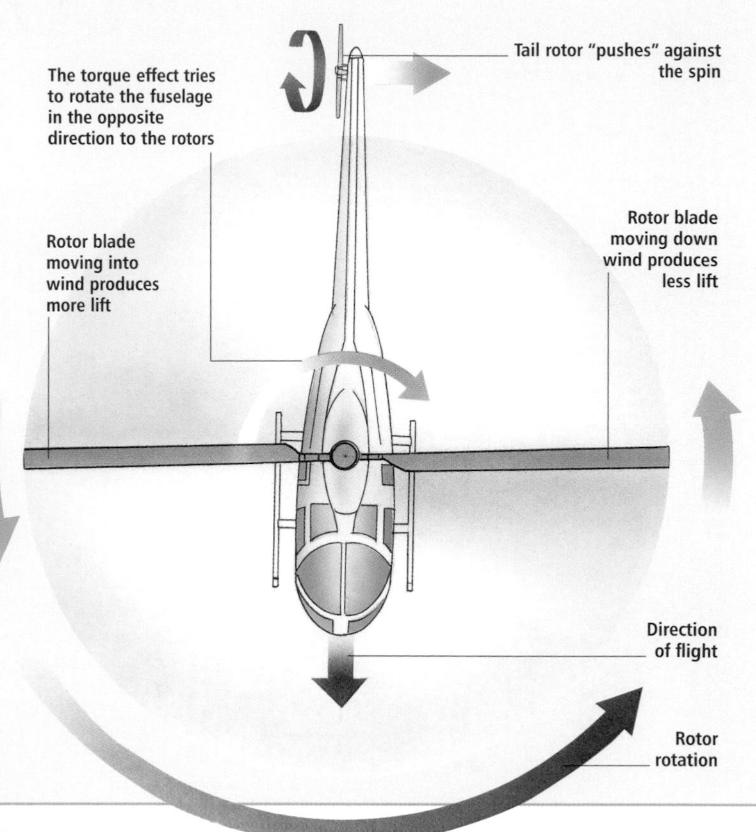

▼ Igor Sikorsky takes the controls of his VS-300 helicopter on a tethered flight in 1939. It was the first successful single-rotor machine and was adopted by both American and British forces in the early years of World War II.

1938–1940 A.D.

ASTRONOMY AND MATH

1938 The asteroid Hermes—first located in 1937 by German astronomer Karl Reinmuth (1892–1979)—approaches within 485,000 miles (780,000 km) of Earth, closer than any other observed asteroid at the time.

1938 American astronomer Seth Nicholson (1891–1963) discovers Lysithea, a small moon of Jupiter.

1938 Russian-born American astrophysicist Otto Struve (1897–1963) establishes the presence of hydrogen ions in space.

1938 German-born American physicist Hans Bethe (1906–2005) and colleagues figure out the series of nuclear fusion reactions that "fuel" stars.

1938 German computer pioneer Konrad Zuse (1910–95) constructs a binary digital computer (the Z1).

1939 Under the pseudonym Nicolas Bourbaki, a group of French mathematicians publishes *Elements of Mathematics,* the first of many books on contemporary mathematics.

CHEMISTRY

1938 American chemist Roy Plunkett (1910–94) synthesizes the "nonstick" plastic PTFE (polytetrafluorethylene, or Teflon).

1939 German chemist Richard Kuhn (1900–67) and coworkers isolate vitamin B_6 (pyridoxine) in yeast and synthesize it within a few months.

1939 French chemist **Marguerite Perey** (1909–75) discovers the radioactive element francium. (She calls it actinium K and changes its name to francium in 1945.)

PHYSICS

1938 German physicists Otto Hahn (1879–1968) and Fritz Strassmann (1902–80) induce nuclear fission in uranium.

1939 French physicists Irène (1897–1956) and Frédéric Joliot-Curie (1900–58) show that uranium fission can lead to a chain reaction.

1939 German-born American physicist **Albert Einstein** (1879–1955) advises President F. D. Roosevelt that the discovery of nuclear chain reactions would lead to the building of bombs.

BIOLOGY AND MEDICINE

Until a live specimen was caught in 1938, the coelacanth was thought to be extinct.

1938 May & Baker pharmaceutical company produces sulfapyridine, the sulfa drug used to treat pneumonia.

1938 American university professor J. L. B. Smith (d. 1968) identifies a coelacanth (a fish previously thought to be extinct).

1938 Physician Florence Seibert (1897–1991) isolates and purifies tuberculin used in the skin test for tuberculosis, first introduced in 1908.

1939 French-born American bacteriologist René Dubos (1901–82) isolates tyrothricin. Tyrothricin (Gramacidin) is the first commercial antibiotic to be marketed.

1939 American zoologist Victor Shelford (1877–1968) introduces the concept of biomes (major geographical areas, such as tundra or desert, that support their own range of organisms).

ENGINEERING AND INVENTION

1938 German-born American engineer Anton Flettner (1885–1961) constructs his Fl-265 "synchropter," a small helicopter with counter-rotating rotors; it first flies in 1939.

1938 The British steam locomotive *Mallard* gains the title of the world's fastest steam engine, with a speed of 126 miles per hour (203 km/h). The record still stands.

1938 American radio engineers Russell (1898–1959) and Sigurd Varian (1901–61) invent the klystron UHF radar transmitter tube.

1938 Italian Achille Gaggia (1895–1961) files a patent for an espresso coffee machine (not manufactured until 1946).

Ferdinand Porsche's design for the German "people's car" (Volkswagen) dates from 1938.

1938 American physicist Katherine Blodgett (1898–1979) creates nonreflective glass.

1938 German engineer Ferdinand Porsche (1875–1951) designs the Volkswagen "Beetle" car.

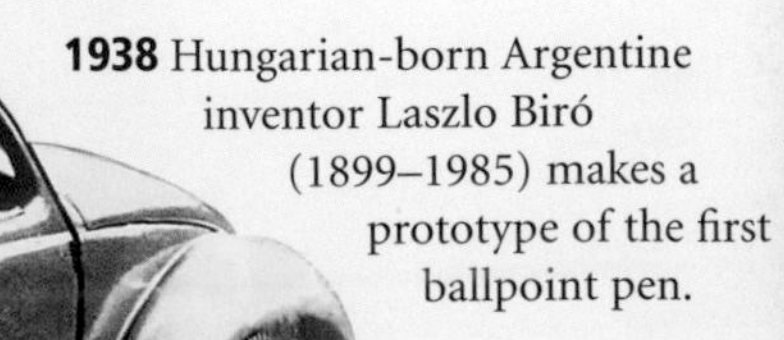

1938 Hungarian-born Argentine inventor Laszlo Biró (1899–1985) makes a prototype of the first ballpoint pen.

1939 American physicist John Atanasoff (1903–95) makes a prototype electronic binary calculator (the Atanasoff-Berry Computer, or ABC).

1939 German-born American geophysicist Walter Elsasser (1904–91) postulates the existence of eddy currents in the Earth's semiliquid outer core to account for the Earth's magnetism.

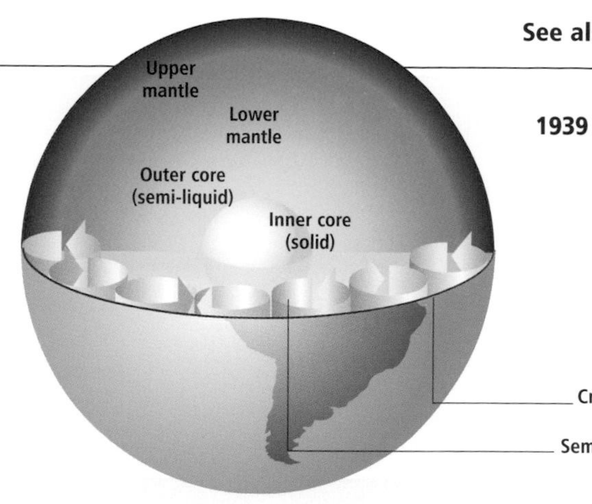

The surface crust of the Earth rides upon layers of molten rock.

1939 Danish astrophysicist Bengt Strömgren (1908–87) identifies Strömgren spheres, regions of ionized gas (mainly hydrogen) surrounding a hot star.

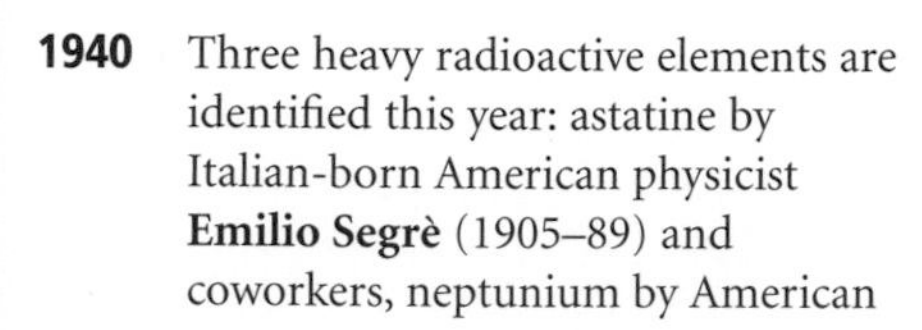

1940 Three heavy radioactive elements are identified this year: astatine by Italian-born American physicist **Emilio Segrè** (1905–89) and coworkers, neptunium by American physicist Edwin McMillan (1907–91) and coworkers, and plutonium by American physical chemist **Glenn Seaborg** (1912–99) and coworkers.

1940 Canadian-born American chemist Martin Kamen (1913–2002) discovers the isotope carbon-14.

1940 American physical chemist Philip Abelson (1913–2004) introduces the diffusion method for separating the isotopes of uranium, an early enrichment process.

1940 Austrian-born American physicist Maurice Goldhaber (1911–) introduces the use of beryllium as a moderator (neutron-absorber) in a nuclear reactor.

Many species of bats—this image shows a leaf-nosed bat—find their way around and locate their prey in the dark using echolocation. Scientists were applying ultrasound in sonar for the underwater detection of submarines when, in 1940, Donald Griffin figured out that bats use a similar system. The bats emit short pulses of ultrasound and detect any echoes that are returned by obstructions or flying insects in their path. Then they home in on their prey silently. In 1960 other scientists discovered that dolphins use a similar echolocation method.

1939 Swiss chemist Paul Müller (1899–1965) determines the insect-killing properties of DDT.

1940 German-born British biochemist Ernst Chain (1906–1979) and Australian pathologist Howard Florey (1898–1968) extract and purify penicillin and perform the first clinical trials of the drug.

1940 Austrian-born American pathologist **Karl Landsteiner** (1868–1943) describes the Rhesus (Rh) factor in human blood.

1940 American zoologist Donald Griffin (1915–2003) announces that bats "echolocate" using ultrasound.

1940 Swiss psychologist Carl Jung (1875–1961) publishes *The Integration of Personality*.

President Franklin D. Roosevelt (left) and shipbuilder Henry Kaiser watch the launch of a 10,500-ton (9,500-tonne) Liberty ship built in just ten days.

1939 English radio engineers John Randall (1905–84) and Henry Boot (1917–83) invent the cavity magnetron UHF radar tube.

1939 The National Broadcasting Company (NBC) begins regular television transmissions in the U. S.

1940 Hundreds of Liberty ships, designed by American engineer William Gibbs (1886–1967), are built to replace ships lost to German submarines.

1940 The Tacoma Narrows suspension bridge in Washington State collapses owing to oscillation (resonant vibration). The phenomenon is explained by Hungarian-born American engineer Theodore Kármán (1881–1963).

1940 German engineer Paul Schmidt invents the pulsejet engine. It is later used to power V-1 flying bombs.

THE DEVELOPMENT OF RADAR

▲ In 1904 Christian Hülsmeyer patented an early-warning system for ships to detect obstructions such as icebergs and other ships. It resembled a form of primitive continuous-wave radar.

During the 1920s and 1930s radio engineers in the United States and Europe reported that passing aircraft distorted their broadcasts. Part of the radio signal had "bounced" off the airplane. Researchers realized that this type of radio reflection could provide a way of detecting planes and other objects such as ships and icebergs.

Radar stands for "radio detecting and ranging," which explains exactly what it does. To detect a plane, a radar set transmits a pulse of very-high-frequency radio (microwaves), and a receiving antenna picks up any radio echoes that return. The direction of the returning signal reveals the direction of the target—all radar echoes are called targets whether they are friend or foe—and its range can be calculated from the time it takes for the microwave signal to travel out and back.

In 1904 German engineer Christian Hülsmeyer (1881–1957) took out the first patents for such a device. He planned a system that used continuous waves (not radio pulses) to warn ships of possible collisions at sea. In 1922 engineers at the U. S. Naval Research Laboratory in Washington, D.C. transmitted radio signals across the Potomac River and detected passing ships when they interrupted the radio beam. In Britain Scottish physicist Robert Watson-Watt (1892–1973) was asked by the Admiralty to investigate the use of radio beams as "death rays" to attack enemy pilots. He found that he could not make a radio signal powerful enough to damage pilots, but he could detect their aircraft. Using the BBC's powerful transmitter in central England, he detected a Heyford bomber flying 7.5 miles (12 km) away at a height of 9,800 feet (3,000 m). He patented his system in 1935. By September 1938, with World War II approaching, the British built a chain of radar antennas on towers 330 feet (100 m) high along the eastern and southern coastline of England. They detected incoming airplanes at a range of up to 200 miles (320 km).

Engineers also adapted radars to aim guns, particularly antiaircraft guns and (in Germany) long-range naval guns. Rudolf Kühnold was Germany's radar pioneer. In 1933 he demonstrated a prototype to the navy in Kiel Harbor. By 1936 several German warships had gunnery radar. In the United States Canadian-born engineer Lawrence Hyland (1897–1989) renewed official interest in antiaircraft and aircraft-detection radars, demonstrating a system on the USS *New York* in 1939.

The very high frequencies of radar signals require special electronics. Early transmitters used a vacuum tube called a magnetron, invented by American physicist Albert Hull (1880–1966) in 1921. An

KEY DATES

1904	Christian Hülsmeyer's patent
1921	Magnetron
1939	Cavity magnetron
1922	U. S. Naval Research Laboratory experiments
1935	Robert Watson-Watt's patent
1938	Klystron
1939	Cavity magnetron
1939	Lawrence Hyland's demonstration

The Cavity Magnetron

The cavity magnetron was invented in 1939. It is a type of radio transmitter tube that produces microwaves and is at the heart of many microwave devices, from microwave ovens to radar sets. It consists of a hollow block of conductive material, such as copper (the anode block). Inside is a heated filament that generates streams of electrons and acts as the cathode. The electrons become concentrated into a negatively charged cloud by the magnetic field between a pair of magnets located above and below the filament. The field also makes the electron cloud move around the filament.

As the cloud passes vanes on the inside of the anode block, the rapidly changing electric charges generate an electromagnetic field that vibrates rapidly within cavities in the anode. An antenna picks up the vibrations, which travel out through a waveguide as microwaves.The cooling fins on the outside of the magnetron help dissipate the extreme heat produced.

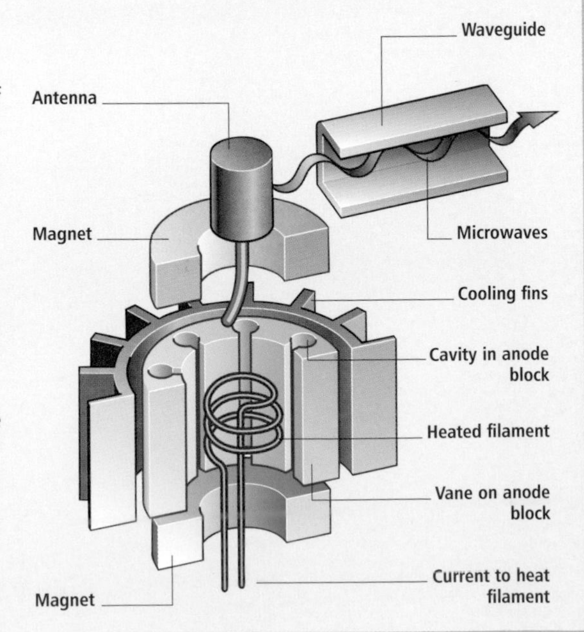

improved version came in about 1934 from the French company CSF, invented by Henri Gutton. The cavity magnetron used a resonant "chamber," or cavity, to create the signals. Two researchers at Birmingham University, John Randall (1905–84) and Henry Boot (1917–83), developed it in 1939. The new device generated wavelengths as small as 3.5 inches (9 cm), and a radar using it could detect a submarine periscope 7 miles (11 km) away. The British government immediately gave details of the cavity magnetron to researchers in the United States. An alternative radar tube, the klystron, was invented in 1938 by American radio engineers Russell Varian (1898–1959) and his brother Sigurd (1901–61).

After World War II radar found more and more peacetime applications. In 1946 astronomers picked up radar signals reflected back from the Moon, and in 1958 echoes came back from the planet Venus. Soviet astronomers made radar contact with Mercury in 1962 and Mars in 1963. The National Aeronautics and Space Administration (NASA) has used orbiting space probes to map the bottom of the Earth's oceans and even the surface of Venus. Weather forecasters make extensive use of satellite radar, and law enforcement agencies employ it to detect speeding motorists.

◀ In the buildup to World War II a series of tall radar towers was put up along the English coast to provide early warning of approaching aircraft. It was known as the Chain Home system.

▼ Radar antennas linked to mobile antiaircraft searchlights proved effective toward the end of World War II in 1945. As soon as the radar "locked on" to an approaching aircraft, the lights were turned on.

ASTRONOMY AND MATH

1941 Russian astronomer Dmitri Maksutov (1896–1964) constructs a new type of reflecting telescope with a wide angle of view.

Maksutov's reflecting telescope of 1941 used a corrector plate to overcome the blurring of the image caused by the inability of a spherical mirror to focus all light from infinity to one focal point. A secondary mirror on the back of the corrector plate reflected light down the tube through a hole in the center of the primary mirror.

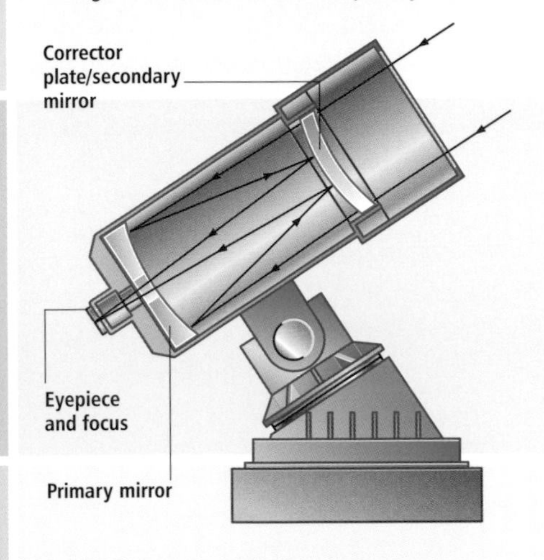

1941 German computer pioneer Konrad Zuse (1910–95) completes his third computer (the Z3). It uses electromechanical relays and, for the first time, punched paper tape.

1941 American mathematician Abraham Albert (1905–72) begins formulating his theories on nonassociative algebra, which he publishes in 1942.

1942 American radio astronomer Grote Reber (1911–2002) compiles the first radio map of the Universe.

1942 English astronomer Harold Spencer Jones (1890–1960) obtains an accurate value for the Earth–Sun distance—equal to 1 astronomical unit, or 93,000,626 miles (149,670,000 km).

Enrico Fermi built the first nuclear reactor at Chicago University and was one of the leaders of the Manhattan Project from 1942.

CHEMISTRY

1941 The German company I.G. Farbenindustrie begins manufacturing polyurethane plastics.

1942 American chemist Louis Fieser (1899–1977) develops napalm for use as an incendiary weapon dropped from the air in bombs.

PHYSICS

1941 Soviet nuclear physicist Georgii Flerov (1913–90) observes spontaneous nuclear fission in uranium.

1942 Italian-American physicist **Enrico Fermi** (1901–54) achieves the first controlled nuclear chain reaction, at the University of Chicago.

BIOLOGY AND MEDICINE

1941 German-born American biochemist Fritz Lipmann (1899–1986) shows the role of ATP (adenosine triphosphate) in the chemistry of cells as the carrier of chemical energy.

1941 American biochemists George Beadle (1903–89) and Edward Tatum (1909–75) observe cell reactions that are controlled by genes.

1941 Canadian-born American surgeon Charles Huggins (1901–97) uses female sex hormones to treat prostate cancer in men.

1941 Ukrainian-born American biochemist Selman Waksman (1888–1973) coins the term "antibiotic" for the new bacteria-killing drugs.

1942 Belgian-born American biologist Albert Claude (1898–1983) first uses an electron microscope in biological studies.

1942 Italian-born American microbiologist Salvador Luria (1912–91) obtains the first high-quality electron micrographs of a bacteriophage (a virus that attacks bacteria).

ENGINEERING AND INVENTION

1941 English engineer Donald Bailey (1901–85) invents the Bailey bridge, a prefabricated metal bridge that can be assembled quickly on the battlefield.

1941 German rocket engineer Walter Dornberger (1895–1980) designs the A-4 liquid-fuel rocket, which first flies in 1942; it is the basis of the V-2 rocket bomb.

1941 The Swiss Federal Railways introduces gas-turbine locomotives.

1941 The prototype of the Gloster Meteor jet fighter plane first flies, with two engines designed by English engineer **Frank Whittle** (1907–96).

1941 English chemists John Whinfield (1901–66) and James Dickson produce the plastic Dacron (Terylene), later licensed to American company DuPont.

1941 French-born American electrical engineer Henri-Gaston Busignies (1905–81) invents a high-frequency direction finder for plotting the positions of aircraft.

1941 Swedish inventor Victor Hasselblad (1906–78) designs the Hasselblad camera. It becomes the top-range, large-format roll-film camera (used by explorers and the American Apollo astronauts).

See also Telescopes **4:**24–25; Rockets **8:**16–17; Nuclear Fission **8:**40–41; The First Computers **8:**44–45

ASTRONOMY AND MATH

1942 English physicist Stanley Hey (1909–2000) discovers that sunspots are a source of ultrashort radio waves.

1942 American physicist John Atanasoff (1903–95) and coworkers complete the development of the ABC, a vacuum-tube electronic calculator with a memory. However, work is interrupted by World War II.

1943 American astronomer Carl Seyfert (1911–60) discovers Seyfert galaxies, which have active nuclei and spectra that indicate a high degree of ionization.

1943 English engineer Thomas Flowers (1905–98) and English logician and mathematician **Alan Turing** (1912–54) build Colossus, an all-electronic stored-program computer, for breaking German military codes.

Early electronic computers made use of banks and banks of vacuum tubes. This 1943 photograph shows Colossus, the first programmable computer, which was built at the British Army's intelligence center to crack German military codes. The Germans believed that their codes were unbreakable; but by deciphering them, the Allies were able to read top-secret messages between Adolf Hitler and his generals, and plan their counteroffensive accordingly.

CHEMISTRY

1942 American biochemist Vincent du Vigneaud (1901–78) establishes the structure of biotin, one of the B complex vitamins.

1943 The American company Dow Corning is established and begins manufacturing silicone plastics.

PHYSICS

1942 Swedish physicist Hannes Alfvén (1908–95) suggests that plasmas such as ionized gases and liquid metals passing through magnetic fields in space create electromagnetic waves.

1943 Japanese theoretical physicist Sin-Itiro Tomonaga (1906–79) publishes a paper on the basic physical principles of quantum electrodynamics.

1943 American scientists complete the world's first operational nuclear reactor at Oak Ridge, Tennessee.

BIOLOGY AND MEDICINE

1943 American biochemist Britton Chance (1913–) establishes the existence of enzyme-substrate complexes in explaining how enzymes work.

1943 Physicians at the University of Stockholm, Sweden, introduce the use of xylocaine as a local anesthetic.

1943 Swiss chemist Albert Hofmann (1906–) accidentally discovers the hallucinogenic properties of the drug LSD (lysergic acid diethylamide).

1943 Dutch-born American physician Willem Kolff (1911–) constructs the first kidney dialysis machine (secretly in German-occupied Netherlands).

The twin-engined Me-262, the first operational jet airplane—nicknamed Schwalbe (Swallow)*—first flew in 1942.*

ENGINEERING AND INVENTION

1942 The Manhattan Project, to make an atom bomb, begins in the U. S.

1942 French underwater explorer Jacques Cousteau (1910–97) invents the aqualung, or scuba (self-contained underwater breathing apparatus).

1942 German rocket engineer ***Wernher von Braun*** (1912–77) designs the German "vengeance weapons," the V-1 flying bomb and the V-2 rocket bomb (both first used in 1944).

1942 The German air force (Luftwaffe) flies the Me-262, the first operational jet airplane, designed by German engineer **Willy Messerschmitt** (1898–1978).

1943 Austrian engineer Paul Eisler (1907–95) makes the first printed circuits for use in products including radio and radar sets, and (much later) computers.

WERNHER VON BRAUN

▲ A young Wernher von Braun studies a model of a two-stage rocket spacecraft. In his later work in the United States he continued to favor the multistage rocket principle.

Wernher von Braun was a trained engineer and pioneer rocket scientist. His career was split into two parts. Until the end of World War II he worked in Germany designing rockets for Hitler's "vengeance weapons." After the war he became an American citizen and worked on NASA's successful space program.

Wernher von Braun (1912–77) was born into a prosperous German family. As a boy he read science fiction by Jules Verne and H. G. Wells, given to him by his mother, an amateur astronomer. Among his favorite books was *The Rocket into Interplanetary Space* by German scientist Herman Oberth (1894–1989). Von Braun attended universities in Berlin and Zurich where he studied engineering. From 1930 he made experimental rockets for the German Society for Space Travel. However, rocket testing was banned in Germany under the terms of the Versailles Treaty (signed after World War I), so he had to continue his work through the military forces. Von Braun fired his rockets from a military base in the Berlin suburb of Kummersdorf. The ballistics and munitions branch of the German army, under German rocket engineer Walter Dornberger (1895–1980), noticed his activities. With the backing of Adolf Hitler (1889–1945) von Braun went to the newly established rocket research center at Peenemünde on the Baltic Sea coast, where in 1936 he became director.

At Peenemünde von Braun's greatest achievement was the "vengeance weapon" 2 (V-2). Designed originally by Dornberger in 1941 as the A-4, the huge rocket used liquid oxygen and alcohol, weighed over 11 tons (10 tonnes), and delivered a warhead containing 1.1 tons (1 tonne) of explosives. Its launch speed of 2,500 feet (760 m) per second carried it high into the upper atmosphere. It came down silently 200 miles (320 km) away at over three times the speed of sound. Many of the 4,300 V-2s launched from 1944 fell on or around Antwerp in Belgium and London.

At the end of the war von Braun and 100 members of his team surrendered to the U. S. Army and went to the United States. Some of the 2,000 V-2 rockets still in storage went with them. From 1946 von Braun worked at White Sands Proving Ground, New Mexico, and in 1950 moved to the Ballistic Missile Agency at Huntsville, Alabama. There he adapted a V-2 to carry a nuclear warhead, creating the Redstone missile. Von Braun became an American citizen in 1955 and was recruited by the National Aeronautics and Space Administration (NASA). In 1958 he oversaw the successful launch of *Explorer I*, the United States's first artificial satellite. He also headed the team that built the Mercury capsules for the U. S. manned spaceflight program, culminating in the orbital flight of John Glenn (1921–) in 1962. In 1960 he became director of the Marshall Space Flight Center, developing the giant three-stage Saturn V rocket for NASA's Apollo missions. The climax of this work came in 1969 when American astronauts landed on the Moon. Von Braun retired from NASA in 1972.

KEY DATES	
1936	Von Braun director of Peenemünde rocket center
1942	Launch of first A-4 (later V-2) rocket
1946	Von Braun works at White Sands in the U. S.
1958	*Explorer I*, first U. S. satellite
1962	Glenn's orbital flight in Mercury capsule
1969	Apollo Moon landing

▲ Jules Verne's fictional rocket to the Moon of 1873 resembled a truncated railroad train. Less than 100 years later real multistage rockets carried Apollo astronauts to the Moon.

▶ The V-2 rocket represented the pinnacle of von Braun's achievements at Peenemünde. Here British troops prepare to test-fire a captured V-2 at the end of World War II.

See also Rockets **8:**16–17; Nuclear Fission **8:**40–41; The Apollo Program **9:**42–43

ASTRONOMY AND MATH

1944 Dutch physicist and radio astronomer Hendrik van de Hulst (1918–2000) predicts (correctly) that interstellar space emits microwave radio waves at a wavelength of 21.1 cm.

1944 German-born American astronomer **Walter Baade** (1893–1960) classifies stars into two types: Population I (younger stars in galaxy arms) and Population II (older stars in galaxy nuclei).

1944 German theoretical physicist and astrophysicist Carl von Weizsäcker (1912–) proposes a theory for the origin of the solar system, a version of the nebular hypothesis first proposed by French astronomer Pierre-Simon de Laplace (1749–1827) in 1796.

1944 American mathematician Howard Aiken (1900–73) and coworkers from IBM complete the Harvard Mark I, an automatic sequence-controlled calculator.

1944 Hungarian-born American mathematician John von Neumann (1903–57) and German-born American mathematician Oskar Morgenstern (1902–76) establish that game theory has a mathematical basis.

CHEMISTRY

1944 English biochemists Archer Martin (1910–2002) and Richard Synge (1914–94) complete the development of their paper chromatography technique; it becomes an important method of chemical analysis.

1944 German biochemist Hans Fischer (1881–1945) determines the structure of the bile pigment bilirubin and synthesizes it.

1944 American chemist **Robert Woodward** (1917–79) and coworkers synthesize the antimalarial drug quinine.

PHYSICS

1945 Soviet physicist Vladimir Veksler (1907–66) designs and builds a powerful particle accelerator, the synchrocyclotron. The first American synchrocyclotron is constructed the following year at Berkeley, California.

1946 American chemist Willard Libby (1908–80) begins work on radiocarbon dating—perfected a year later—which estimates the age of old organic material from the amount of radioactive carbon-14 isotope it still contains.

1946 American experimental physicist **Luis Alvarez** (1911–88) designs a proton linear accelerator.

1946 The first Soviet nuclear reactor goes into service, designed by Soviet nuclear physicist Igor Kurchatov (1903–60).

BIOLOGY AND MEDICINE

1944 Canadian-born American bacteriologist Oswald Avery (1877–1955) and coworkers demonstrate that nearly all organisms have DNA (deoxyribonucleic acid) as their hereditary material.

1944 Ukrainian-born American biochemist Selman Waksman (1888–1973) discovers the bacteria-killing drug streptomycin.

1944 American pediatrician Helen Taussig (1898–1986) and American surgeon Alfred Blalock (1899–1964) introduce a pulmonary bypass operation to treat the heart defect in "blue babies."

1945 Chinese-born American biochemist Choh Hao Li (1913–87) isolates the human growth hormone somatotropin.

1945 American biochemist **Melvin Calvin** (1911–97) begins his studies of photosynthesis using radioactive carbon-14 as a tracer to follow the reactions.

ENGINEERING AND INVENTION

1944 The Delaware Aqueduct, part of New York City's water-supply system, is completed. It includes an 85-mile (137-km) tunnel, later extended to 105 miles (169 km)—at the time the world's longest tunnel.

1944 The German air force employs the unsuccessful Messerschmitt Me-163B Komet rocket-powered airplanes as interceptors.

1944 German forces begin launching V-1 flying bombs and V-2 rocket bombs against southeastern England.

1944 Russian-born American engineer Igor Sikorsky (1889–1972) sets the design of the modern helicopter with the VS-36A: an adjustable-pitch main rotor and a single vertical tail rotor for stability.

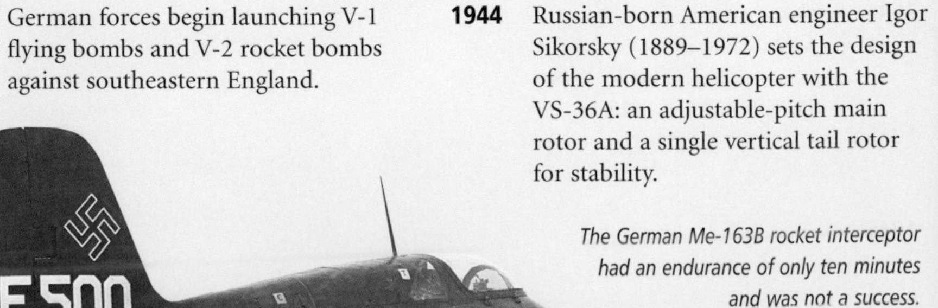

The German Me-163B rocket interceptor had an endurance of only ten minutes and was not a success.

Electronic counters from the Harvard Mark 1 digital computer of 1944.

1944 American pioneer computer engineers **John Eckert** (1919–95) and **John Mauchly** (1907–80) devise the mercury delay line store (a computer memory device). Two years later they complete ENIAC (Electronic Numerical Integrator and Computer), a fully electronic computer.

1946 English physicist Edward Appleton (1892–1965) detects radio emissions from sunspots.

1946 Hungarian physicist Zoltán Bay (1900–92) and investigators at the U. S. Army Signal Corps Laboratory independently obtain radar reflections from the Moon.

1946 English engineer Frederic Williams (1911–77) makes a computer memory using a cathode-ray tube.

1944 American physical chemist **Glenn Seaborg** (1912–99) and coworkers isolate the radioactive elements americium and curium.

1945 English crystallographer **Dorothy Hodgkin** (1910–94) uses X-ray crystallography to determine the structure of penicillin.

1946 English chemist Robert Robinson (1886–1975) figures out the structure of the alkaloid drug strychnine.

Nuclear magnetic resonance (NMR) was developed by Felix Bloch in 1946. It involves exposing a chemical in solution simultaneously to high-frequency radio radiation in a surrounding coil and a very strong magnetic field provided by powerful magnets. A second coil detects the radio frequencies that are absorbed by the chemical. They are analyzed in a signal detector and plotted as a spectrum. In this example the NMR spectrum distinguishes between the hydrogen atoms (H) in two different chemical groupings within the chemical bromoethane (CH_3-CH_2-Br). The peaks on the chart correspond to the CH_3 and CH_2 groups.

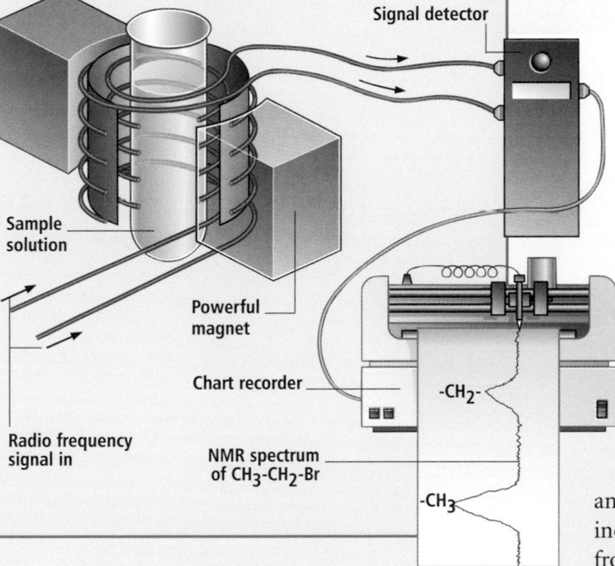

1946 Swiss born American physicist Felix Bloch (1905–83) and American physicist Edward Purcell (1912–97) develop the technique of NMR (nuclear magnetic resonance) spectroscopy, which becomes a powerful tool in chemical analysis.

1946 Swedish pharmacologist Ulf von Euler (1905–83) identifies the hormone norepinephrine (noradrenaline).

1946 British surgeon Thomas Cecil Gray introduces the use of curare (an American-Indian arrow poison) in general anesthesia.

1946 In the U. S., German-born biologist Max Delbrück (1906–81) and biologist Alfred Hershey (1908–97) independently find that genetic material from two different viruses can combine to form a new virus.

1946 French–born American bacteriologist René Dubos (1901–82) finds a way to culture the tuberculosis bacillus.

1944 American physicist Robert Dicke (1916–97) makes a radiometer for detecting microwave radiation.

1945 American government scientists make and test an atom bomb.

1945 Engineers working for the British Rolls-Royce company develop the afterburner for jet engines.

1945 Austrian-born American electrical engineer **Rudolf Kompfner** (1909–77) invents the traveling-wave amplifier for microwave signals, such as those used in radar.

1945 Northern Irish engineer James Martin (1893–1981) designs the ejector seat for aircraft, which will save the lives of thousands of pilots.

1946 American engineer Percy Spencer (1894–1970) invents the microwave oven while working for the Raytheon Company.

1946 The Japanese company Sony is founded (originally as Tokyo Tsushin Kogyo) by Japanese engineer Akio Morita (1921–99).

NUCLEAR FISSION

▲ John Cockcroft, photographed in 1932 with the particle accelerator he and Ernest Walton had designed at the Cavendish Laboratory, Cambridge. In recognition of their work the two men shared the 1951 Nobel Prize for Physics.

In the early 20th century physicists were learning what happens when atoms are bombarded with subatomic particles. Their experiments led them to realize that under certain circumstances such bombardment could release huge amounts of energy through nuclear fission in nuclear reactors, which could be used to generate electricity. In January 2005 there were 439 operational reactors worldwide, producing 16 percent of all electricity.

In 1932 English physicist John Cockcroft (1897–1967) and Irish physicist Ernest Walton (1903–95) began experimenting with high-energy protons in the particle accelerator they had built in Cambridge, England. In Paris in 1934 French physicists Irène (1897–1956) and Frédéric (1900–58) Joliot-Curie found that proton bombardment sometimes produced radioactive isotopes of the target atoms. Two years later Italian-American physicist **Enrico Fermi** (1901–54), working in Rome, found that neutrons—discovered in 1932 by English physicist James Chadwick (1891–1974)—were more effective than protons at smashing atoms.

Neutron bombardment usually produced heavier atoms through neutron absorption; but when Fermi bombarded atoms of certain heavy elements, especially uranium, he found that much lighter nuclei were produced. In 1939 German physicists Otto Hahn (1879–1968) and Fritz Strassmann (1902–80) identified the products of uranium bombardment as elements with about half the mass of uranium. Hahn and Strassmann had demonstrated that the uranium nuclei had broken apart. Fission had occurred.

The same year Lise Meitner (1878–1968), an Austrian-born physicist working in Stockholm, and her Austrian nephew Otto Frisch (1904–79), working in Copenhagen with Danish physicist **Niels Bohr** (1885–1962), explained this result. The uranium nucleus absorbed a neutron, causing it to vibrate violently and then divide into two parts with the release of about 200 million electron volts of energy ($= 3.204 \times 10^{-11}$ joules). Hahn and Strassmann then found that in addition to a large amount of energy, uranium fission released neutrons that could trigger fission in other uranium nuclei, raising the possibility of a chain reaction that would release a prodigious amount of energy. The Joliot-Curies and Leo Szilard confirmed this experimentally. Szilard (1898–1964) was a Hungarian American physicist working with Fermi, then at Columbia University, New York.

Three isotopes of uranium occur naturally, always in the same proportions: uranium-238 (^{238}U) accounts for 99.28 percent, ^{235}U for 0.71 percent, and ^{234}U for 0.006 percent. Bohr calculated that fission would occur more readily in ^{235}U than in either of the other isotopes. This meant that a way had to be found to separate the isotopes—a technique now known as "enrichment." Bohr also figured out that fission would be more effective if the neutrons were slowed down. Szilard and Fermi suggested surrounding the uranium with a "moderator"—a substance such as graphite or heavy water that slows neutrons.

Just two days before the outbreak of war in 1939 Bohr and American theoretical physicist John Wheeler (1911–) published a paper describing the complete fission process. Also in 1939 French physicist Francis Perrin (1901–92) showed that a certain "critical mass" of uranium is needed to sustain a chain reaction by ensuring that enough of the neutrons released strike other uranium nuclei. Perrin also introduced the idea of adding a substance that absorbs (rather than slows) neutrons as a way to control the rate of the fission reaction. Rudolf Peierls (1907–95), a German-born physicist working in England, extended these ideas. Designed by Fermi and called "an atomic pile," the first working reactor began operating on December 2, 1942, at the University of Chicago. In 1951 the Experimental Breeder Reactor, built at Idaho National Engineering Laboratory near Idaho Falls, Idaho, became the first reactor to generate electricity.

It became apparent to scientists that a sustained fission reaction could be used to create a bomb of immense power. Work to develop an atom bomb began in Britain and the United States. The two programs were merged in August 1942 to form the Manhattan Project. The first successful test took place in New Mexico on July 16, 1945.

Research in the Soviet Union had been proceeding independently, and by 1940 Soviet scientists also understood the principles of fission and recognized the possibility of a chain reaction. But it was 1942 before Stalin became convinced that a bomb could be developed, and a program led by nuclear physicist Igor Kurchatov (1903–60) began. The first Soviet reactor entered operation in 1948, and the first Soviet bomb was detonated in August 1949.

KEY DATES	
1932	Cockcroft and Walton particle accelerator
1934	Joliot-Curies produce radioactive isotopes
1936	Fermi uses neutrons to smash atoms
1939	Hahn and Strassman identify products of uranium fission
1942	First nuclear reactor
1945	First atom bomb tested
1951	Nuclear reactor built to generate electricity

Uranium Fission

Fission occurs when an atomic nucleus breaks into two parts—"fission products"—with the release of two or three neutrons. Some heavy elements undergo fission spontaneously. In other elements fission is induced by bombarding the nucleus with neutrons or protons. Fission releases an amount of energy equal to the energy that used to bind the nucleons together.

When a slow-moving neutron strikes a ^{235}U atom, the nucleus absorbs the neutron, becoming ^{236}U, which immediately divides into two lighter nuclei with the release of neutrons. If the neutrons strike other ^{235}U nuclei, the latter also undergo fission, setting up a chain reaction. The fission of 2.2 pounds (1 kg) of ^{235}U nuclei releases 20,000 megawatt-hours of energy, equivalent to burning 3.3 million tons (3 million tonnes) of coal.

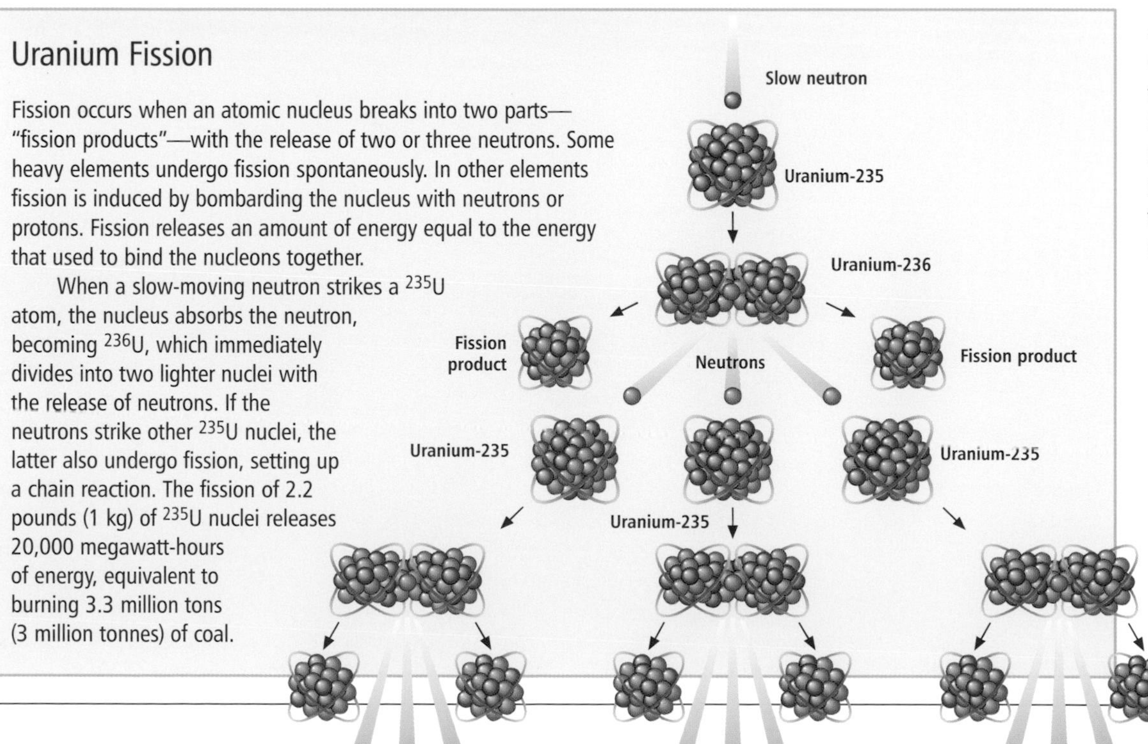

▲ Scientists observe a controlled chain reaction in the atomic reactor at the University of Chicago in 1942. Because of the radiation no photographs could be taken, so the event was recorded by an artist.

ASTRONOMY AND MATH

1947 Armenian astronomer Viktor Ambartsumian (1908–96) describes stellar associations, groups of stars in the galaxy's spiral arms (but not as closely linked as a star cluster).

1948 English astronomer Fred Hoyle (1915–2001) and Austrian-born English astronomers Hermann Bondi (1919–2005) and Thomas Gold (1920–2004) propose the steady-state theory of the Universe (i.e., that it is not expanding—it maintains a constant average density).

1948 Dutch-born American astronomer Gerard Kuiper (1905–73) locates Miranda, a small inner moon of Uranus. A year later he finds Nereid, Neptune's moon.

1948 American physicist Ralph Alpher (1921–), German-born American Hans Bethe (1906–2005), and Russian-born American **George Gamow** (1904–68) propose the alpha, beta, gamma theory for the origin of the Universe, which corresponds to the big bang theory.

1948 American mathematician Norbert Wiener (1894–1964) coins the term "cybernetics" in his book *Cybernetics: or Control and Communication in the Animal and the Machine.*

1948 American astronomers Harold (1882–1968) and Horace (1912–2003) Babcock detect the Sun's magnetic field.

1948 American mathematician Claude Shannon (1916–2001) formulates information theory.

CHEMISTRY

1947 American chemist Jacob Marinsky (1918–2005) and coworkers bombard neodymium with neutrons and confirm the production in 1945 of the radioactive element promethium.

1948 Scottish biochemist Alexander Todd (1907–97) synthesizes ADP and ATP (adenosine diphosphate and adenosine triphosphate).

1948 American organic chemist Karl Folkers (1906–97) determines the structure of the bacteria-killing drug streptomycin.

PHYSICS

1947 American physicist Willis Lamb (1913–) observes the Lamb shift, a small difference in energy between two levels in the atomic spectrum of hydrogen; it contributes to quantum electrodynamic theory.

1947 English physicist Cecil Powell (1903–69) and Italian physicist Giuseppe Occhialini (1907–93) discover the pion (pi-meson), a heavy subatomic particle, in cosmic rays.

1948 Hungarian-born English electrical engineer Dennis Gabor (1900–79) puts forward the principles of holography (although their application in hologram production has to await the development of lasers).

BIOLOGY AND MEDICINE

1947 German-born American biochemist Fritz Lipmann (1899–1986) isolates coenzyme A (CoA), a substance needed to ensure the effectiveness of certain enzymes.

1947 Microbiologist Mildred Rebstock discovers the antibiotic chloramphenicol (Chloromycetin) in a sample of soil from Venezuela.

1947 Italian biologist Rita Levi-Montalcini (1909–) finds nerve growth factor (NGF) in chick embryos.

1947 American geneticist Joshua Lederberg (1925–) reports that some bacteria can reproduce by conjugation (joining together as in sexual reproduction).

1948 American biochemists Philip Hench (1896–1965) and Edward Kendall (1886–1972) introduce the use of cortisone to treat patients with rheumatoid arthritis.

1948 English anthropologist **Mary Leakey** (1913–96) discovers fossils of the humanoid *Proconsul africanus* in Africa.

ENGINEERING AND INVENTION

1947 American airman Charles "Chuck" Yeager (1923–) makes the first supersonic flight, in a Bell X-1 rocket-propelled airplane.

Chuck Yeager wriggles out of the Bell X-1 airplane after its first supersonic flight.

1947 American inventor Edwin Land (1909–91) demonstrates the Polaroid instant camera.

1947 American architect **Richard Buckminster Fuller** (1895–1983) devises the geodesic dome construction for large buildings.

1947 The first nuclear reactor in Europe is built at Harwell, England.

1947 American physicists **John Bardeen** (1908–91) and **Walter Brattain** (1902–87) and English-born American **William Shockley** (1910–89) build the point-contact transistor. A year later Shockley conceives the junction transistor.

1947 Soviet gunsmith Mikhail Kalashnikov (1919–) produces the AK-47 assault rifle (used by Soviet forces from 1949).

See also The Elusive Electron **7**:16–17; Rockets **8**:16–17; The First Computers **8**:44–45; New Chemical Elements **9**:16–17

ASTRONOMY AND MATH

1949 German-born American astronomer **Walter Baade** (1893–1960) discovers the asteroid Icarus, which passes very close to the Sun (and the Earth).

1949 American pioneer computer engineers **John Eckert** (1919–95) and **John Mauchly** (1907–80) construct BINAC, a binary automatic computer.

1949 English computer engineer Maurice Wilkes (1913–) and coworkers build EDSAC, an electronic delay-storage automatic calculator.

1949 The American company IBM makes a stored-program electromechanical computer SSEC (selective sequence electronic calculator); in the same year workers at Manchester University, England, make a stored-program electronic computer (Manchester University Mark I).

1949 American engineer Jay Forrester (1918–) produces a magnetic core memory for a computer.

1949 American astronomer Fred Whipple (1906–2004) proposes the "dirty snowball" theory of comets, suggesting that they consist mainly of ice with accumulated rocky debris.

Fred Whipple likened the nucleus of a comet to a dirty snowball and suggested that it consists of a collection of ice and small solid particles, together extending a few miles across. A bright coma develops around the nucleus as it evaporates on approaching the Sun. There may be one or two tails pointing away from the Sun, one made of dust or gas and one consisting of ions.

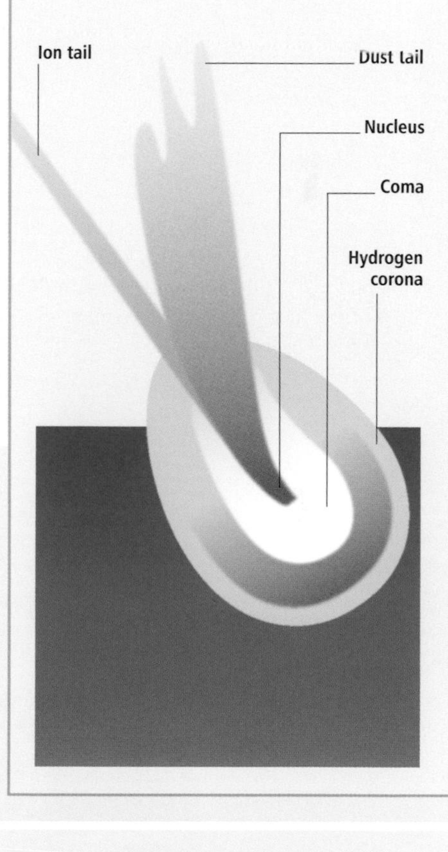

CHEMISTRY

1949 English chemists George Porter (1920–2002) and Ronald Norrish (1897–1978) study ultrafast chemical reactions, using pulses of light in the technique known as flash photolysis.

1949 American physical chemist **Glenn Seaborg** (1912–99) and team produce the radioactive element berkelium by bombarding americium with alpha particles (helium nuclei).

PHYSICS

1948 American theoretical physicist Richard Feynman (1918–88) and coworkers and, independently, American physicist Julian Schwinger (1918–94) formulate new versions of quantum electrodynamics.

1949 German-born American physicist Maria Goeppert-Mayer (1906–72) proposes the "shell" theory of the atomic nucleus (which pictures its component protons and neutrons moving in shells analagous to those occupied by electrons).

BIOLOGY AND MEDICINE

1948 American organic chemist Karl Folkers (1906–97) and coworkers isolate vitamin B_{12} (cyanocobalamin), the factor that prevents pernicious anemia.

1948 The World Health Organization (WHO) is founded.

1949 American microbiologists John Enders (1897–1985) and Frederick Robbins (1916–2003) culture the virus that causes poliomyelitis.

1949 Canadian geneticist Murray Barr (1908–95) identifies Barr bodies, condensed X chromosomes in the cells of female mammals (used in sex-determination tests).

ENGINEERING AND INVENTION

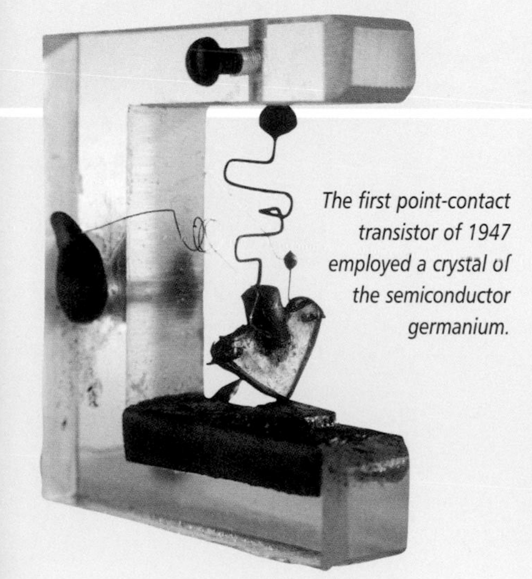

The first point-contact transistor of 1947 employed a crystal of the semiconductor germanium.

1948 American electrical engineer Peter Goldmark (1906–77) invents the long-playing phonograph record.

1948 American instrument maker Leo Fender (1909–91) and coworkers market the first solid-body electric guitar.

1948 Swiss-born Belgian physicist August Piccard (1884–1962) designs the bathyscaphe submersible craft.

1948 The first turboprop airliner, the Vickers Viscount, makes its maiden flight in England.

1949 The first jet airliner, the De Havilland Comet, flies in England.

1949 American rocket engineers launch a two-stage rocket (a German V-2 with a small rocket mounted on its nose) at White Sands, New Mexico.

THE FIRST COMPUTERS

KACHYESTVO UGLYA OPRYEDYELY AYETSYA KALORYIYNOSTJYU

This card is punched with a sample Russian language sentence (as interpreted at the top) in standard IBM punched-card code. It is then accepted by the 701, converted into its own binary language and translated by means of stored dictionary and operational syntactical programs into the English language equivalent which is then printed.

▲ This punched card was used on an IBM 701 computer to translate instantly a sentence from Russian into readable English. The Russian sentence (at the top of the card) translated as: "The quality of coal is determined by calorie content."

A computer is an electronic machine that can perform various tasks involving the processing of information or data under instructions from a program. Today the term most often applies to a digital computer, which handles its data in the form of digits or numbers expressed in binary notation.

Binary is a number system that uses only two digits, 1 and 0, represented in the computer's activities and memory as on and off pulses of electric current. According to the definition above, the first computers were used by the U. S. Army and Navy toward the end of World War II. They were massive vacuum-tube machines, developed from electronic calculating machines of the late 1930s, which in turn derived from earlier mechanical calculating machines. The first calculator was the abacus, a frame with beads invented in about 3,000 B.C. and still used in parts of China and Japan. The task of multiplication is simplified by logarithms, invented in 1614 by Scottish mathematician **John Napier** (1550–1617) and "mechanized" in the slide rule designed in 1622 by English mathematician William Oughtred (1574–1660).

French scientist **Blaise Pascal** (1623–62) probably invented the first mechanical adding machine in 1642. It had a system of intermeshed cogs, a method also adopted by English mathematician Charles Babbage (1792–1871) in his "analytical engine" of 1833. Babbage's machine could be programmed for a particular calculation and was therefore a computer (although not electronic). Calculating machines with keyboards—called comptometers—were developed from the 1880s by inventors such as the American William Burroughs (1855–98). Later versions of such machines provided a printout of the results.

Most early methods of feeding data into programmable machines used punched tape or punched cards. In about 1805 French inventor Joseph Jacquard (1752–1834) designed a loom that could weave various patterns in carpets by following instructions on an endless belt of punched cards. American inventor Herman Hollerith (1860–1929) employed similar cards to record and analyze the results of the 1890 U. S. census. The company that Hollerith formed later became part of International Business Machines (IBM).

Electromechanical calculators appeared in the 1930s, invented by American scientists such as Vannevar Bush (1890–1974) and John Atanasoff (1903–95). By 1942 Atanasoff constructed an electronic calculator (the ABC), using vacuum tubes, that could also be programmed for processing data—arguably the first true computer. Two years later at Harvard, American mathematician Howard Aiken (1900–73) had a manually operated digital machine controlled by punched paper tape, and in 1946 the all-electronic ENIAC (Electronic Numerical Integrator and Calculator) computer was in service, although it too still employed vacuum tubes.

KEY DATES

1833	Babbage's "Analytical Engine"
1890	Hollerith punched cards
1942	Electronic calculator
1946	ENIAC and Neumann's stored-program computer
1951	First mass-produced computer (UNIVAC I)

▶ The stored-program computer built at Manchester University in 1949 occupied a whole laboratory. Despite its size, it had far less computing power than a modern laptop.

▲ Magnetic tape improved the speed and storage capacity of computers. A row of tape storage units can be seen in front of the UNIVAC computer used in Washington, D.C., to tabulate the results of the 1954 U. S. census.

◀ Computer pioneer John Mauchly compares an abacus counting frame with a bank of switches on the ENIAC general-purpose computer of 1946. The computer had to be reprogrammed by hand for each application.

A machine built at Princeton in 1946 by Hungarian-born American mathematician John von Neumann (1903–57) was the first with a stored program that used binary numbers. The idea was incorporated into UNIVAC I, designed by American computer engineers **John Eckert** (1919–95) and **John Mauchly** (1907–80) and in 1951 the first computer to be manufactured in quantity in the United States. They modified it to use magnetic tape storage a year later. In 1949 a team at Manchester University, England, also built a stored-program machine under the guidance of English mathematician **Alan Turing** (1912–54), who had worked at Princeton. The Manchester machine was so successful that the British government commissioned the Ferranti company to manufacture it. In a few years it sold eight of the Mark 1 computers—a large number at the time.

After the invention of the transistor in the late 1940s by physicists in the United States, computers became faster and smaller. By the mid-1960s silicon chips had arrived, so that circuits designed in 1970 incorporated a complete computer microprocessor on a single chip. Today microchips are used not only in personal computers (PCs) but also in embedded systems in domestic appliances, automobiles, and industrial robots.

KEY PEOPLE 1920–1949 A.D.

People whose names appear in **bold type** *have their own articles in this section or in the "Key People" section of another volume. Names in* ***bold italics*** *indicate they are the subject of a special feature.*

Aston, Francis (1877–1945)

Francis William Aston was an English physicist who developed the mass spectrograph that separates and identifies atoms, particularly isotopes of the same element. He was born near Birmingham in central England and educated at Mason College (later Birmingham University) and Cambridge, where he studied chemistry. From 1898 to 1900 he worked with English chemist Edward Frankland (1825–99) on optical activity. Aston's expertise with high vacuums led physicist John Poynting (1852–1914) to invite him back to Birmingham University. He stayed there until 1910, when he went to the Cavendish Laboratory at Cambridge under English physicist **J. J. Thomson** (1856–1940). Apart from spending some time at the Royal Aircraft Establishment in the south of England during World War I, he remained at Cambridge for the rest of his career.

Aston's mass spectrograph is basically a discharge tube in which positive rays (positively charged ions) from the anode pass through a magnetic field and an electric field so that they are deflected. The amount of deflection depends on their masses, and the ions become separated (the lighter ones being deflected less) before they register their arrival on a photographic plate or film.

The rare gas neon was Aston's first subject for study in 1919. He discovered that the element consists of two isotopes of masses 20 and 22 in the proportion of 10 to 1, which accounted for neon's nonintegral atomic mass of 20.2. Likewise, he was able to account for the "odd" atomic mass of chlorine (35.5): It consists of a mixture of isotopes of masses 35 (75 percent) and 37 (25 percent). Aston went on to use the instrument to determine accurately the atomic masses of 212 of the naturally occurring isotopes known at the time.

Baade, Walter (1893–1960)

Wilhelm Heinrich Walter Baade was a German-born American astronomer whose studies of the stars in the Milky Way led to discoveries about the size of the Universe. He was born in Schröttinghausen, the son of a schoolteacher, and went to college in Münster and Göttingen. He worked at the Hamburg Observatory from 1919 before moving in 1931 to the United States. He joined the staff of the Mount Wilson Observatory in California, where he remained until he retired in 1958.

Baade's first notable discovery came in 1920, when he found the asteroid Hidalgo, which orbits as far out as Jupiter. At the other extreme the orbit of the asteroid Icarus, which he found in 1949, lies entirely within that of Mercury.

In 1942 Baade turned his telescope on the Andromeda galaxy and was able to show that the "blur" at the center actually consists of separate stars. He classified these older, redder stars in 1944 as Population II stars, in contrast to the younger blue Population I stars in the spiral arms of the galaxy. Starting in 1951 he went on to investigate the newly discovered radio sources in space to see if any of them coincided with a visible "optical" object.

Baird, John Logie (1888–1946) *See* **feature Vol. 8, pages 12–13.**

Banting, Frederick (1891–1941)

Frederick Grant Banting was a Canadian physiologist who, with Canadian Charles Best (1899–1978), discovered the hormone insulin and pioneered its use to treat patients with diabetes. Banting was born in Alliston, Ontario, the son of a farmer. From 1910 he studied medicine at Toronto University and served in the Canadian Army Medical Corps during World War I. He won a medal for gallantry and left the army on medical grounds. After the war he worked as a demonstrator at Western University Medical School. Using borrowed laboratory space, in 1921 he and Best found a new method of preparing an extract of the pancreas containing insulin for treating diabetes. By 1922 they had prepared insulin that was pure enough for clinical trials.

In 1923 Banting and Best, together with Canadian biochemist James Collip (1892–1965), patented the hormone, and in the same year Banting was appointed professor of medical research at Toronto. He joined an army medical unit at the outbreak of World War II in 1939, researching mustard gas and problems with blackouts (sudden unconsciousness) in aircrew. He died when his airplane crashed in the snow on a flight from Gander, Newfoundland, to England.

Braun, Wernher von (1912–77) *See* **feature Vol. 8, pages 36–37.**

de Broglie, Louis (1892–1987)

Louis-Victor, Duc de Broglie, was a French physicist who put forward the theory that particles can sometimes behave as if they were waves—the so-called wave-particle duality. He was born in Dieppe in northern France into a noble family and was educated at the Sorbonne in Paris. He trained as a historian but became interested in science during World War I when, as a member of a wireless telegraphy unit, he was posted to a station at the Eiffel Tower. He then took up physics and received a doctorate from the Sorbonne in 1924. From 1926 he taught there, moving two years later as professor of physics to the new Henri Poincaré Institute, where he remained for 34 years.

In 1905 German-born American physicist **Albert Einstein** (1879–1955) had explained the photoelectric effect (in which light strikes a metal surface and releases electrons). He said that the light (a radiation wave) consists of individual quanta, called photons, that interact with the electrons in the metal as discrete particles. De Broglie postulated in 1924 that particles can also behave like waves. The idea was proved experimentally in 1927 and employed two years later by Austrian physicist Erwin Schrödinger (1887–1961) in his new wave mechanics.

Carlson, Chester (1906–68)

Chester Floyd Carlson was an American physicist who invented the photocopying process called xerography (meaning "dry writing") in 1937. He trained as a physicist at the California Institute of Technology and in 1930 went to work for Bell Telephone Laboratories. Because his work involved patents, he qualified as a patent attorney in 1934 and joined an electrical company. He experimented with ways of copying documents. In 1940 he patented his xerographic process, which relies on the fact that some substances, such as selenium, lose any electric charge when light shines on them. The Haloid company, which made photographic materials, took up the idea in 1946 and four years later marketed the first Xerox photocopier. Carlson made a fortune from royalties, much of which he used for charitable purposes in his later years.

Compton, Arthur (1892–1962)

Arthur Holly Compton was an American physicist best known for his discovery and explanation of the change in the wavelength of X-rays when they collide with electrons in metals (the Compton effect). He was born in Wooster, Ohio. He studied at Wooster College, Princeton, and Cambridge Unversity, England, where he received his doctorate in 1916. Later he worked for the Westinghouse Electric Company. In 1920 he became professor of physics at Washington University, St. Louis, and then moved to the University of Chicago in 1923. At Washington University and later he measured the scattering of X-rays by various light elements, such as carbon, and found that the scattered X-rays underwent a shift to a longer wavelength, which he put down to their interaction with free electrons in the material. The Compton effect provides evidence that electromagnetic radiation (such as X-rays and light) consists of particles, or quanta. In 1938 he showed that cosmic rays consist of charged particles.

Compton joined the team on the Manhattan Project to make the first atom bomb and in 1942 worked with **Enrico Fermi** (1901–54) at Chicago, building the first nuclear reactor. He returned to Washington University in 1945 as chancellor and from 1953 to 1961 was professor of natural philosophy there.

Fermi, Enrico (1901–54)

Enrico Fermi was an Italian-born American physicist who undertook major work in nuclear physics from 1925 to 1950. He was born in Rome, son of a government official. From 1918 he was educated at the Scuola Normale Superiore in Pisa and then attended the university there, receiving his doctorate in 1922. He spent some time at universities in Germany and the Netherlands, and in 1927 became professor of theoretical physics at Rome University.

In 1934, by placing a slab of paraffin wax in a beam of fast neutrons, he generated slow neutrons, which he used to bombard various elements in order to produce new artificial radioisotopes. He also worked out a theory of beta decay, showing that in a radioactive substance a neutron changes into a proton, an electron (beta particle), and a neutrino. The existence of the latter had been predicted in 1930 by Austrian-born Swiss physicist **Wolfgang Pauli** (1900–58).

By 1938 growing anti-Semitism in Italy forced Fermi to emigrate to the United States (his wife was Jewish). In 1939 he was appointed professor of physics at Columbia University. On a disused squash court at the University of Chicago he constructed the first atomic pile (nuclear reactor) in 1942, using several tons of enriched uranium as the fuel and 40,000 blocks of graphite as the moderator to slow down neutrons. In 1945 he witnessed the testing of the first atom bomb in New Mexico. A tiny unit of length (equal to 10^{-15} meters) is named the fermi in his honor, and in 1953 a team of American chemists named the newly discovered element 100 fermium.

Fessenden, Reginald (1866–1932)

Reginald Aubrey Fessenden was a Canadian-born American radio engineer who made pioneering inventions during the early years of radio. He was born in East Boston, Quebec. After attending school in Ontario and graduating from Bishops College, Lennoxville, Quebec, he became principal of the Whitney Institute in Bermuda in 1885. A year later he went to work as chief chemist at the New Jersey research laboratories of American engineer **Thomas Edison** (1847–1931). In 1890 he moved to the Westinghouse Electric Company in Pittsburgh and in 1892 took an academic appointment at Purdue University, Lafayette, Indiana. A year later he became professor of electrical engineering at the University of Pittsburgh, where he stayed until 1900. After two years with the U. S. Weather Bureau he founded the National Electric Signaling Company.

On December 24, 1906, Fessenden made the first American radio broadcast from his transmitter at Brant Rock, Massachusetts. It employed a new high-frequency alternator and his invention of amplitude modulation (AM), in which the broadcast signal is made to modulate (vary) the amplitude of a continuous carrier wave. Ships several hundred miles away picked up the signals. Among Fessenden's other many inventions were the heterodyne circuit, which greatly improves the quality of radio reception, a depth finder that uses sound waves (an echosounder), a radio compass, and signaling devices used in submarines.

Fleming, Alexander (1881–1955)

Alexander Fleming, a Scottish bacteriologist, was the discoverer of penicillin. He was born in southwest Scotland, the son of a farmer. At the age of 16 he took a job as a shipping clerk in London. In 1902 he enrolled as a medical student at St. Mary's Medical School. He graduated in 1906, undertook research for two years, and worked as a lecturer. Apart from serving in the Royal Army Medical Corps during World War I, he worked at St. Mary's for his entire career.

Fleming was the first person to use the vaccine against typhoid fever, discovered in 1892 by English bacteriologist Almroth Wright (1861–1947). He also introduced into Britain the use of Salvarsan, discovered in 1910 by German chemist Paul Ehrlich (1854–1915), to treat syphilis. Fleming was interested in controlling infection in wounds and studied the effects of body secretions on bacteria. That was how in

1921 he found lysozyme, an enzyme present in saliva, serum, and tears, which destroys bacteria. He made his best-known discovery (penicillin) in 1928. He accidentally left a culture dish of staphylococcus bacteria uncovered and noticed some clear areas where the bacteria had been killed by a substance produced by a contaminating mold. He identified the mold as *Penicillium notatum* and called the bacteria-killing substance penicillin. He found that it also killed some other (but not all) bacteria. The substance itself was isolated in 1940 by Australian pathologist Howard Florey (1898–1968) and German biochemist Ernst Chain (1906–79).

Gamow, George (1904–68)

George Gamow was a Russian-born American physicist who made discoveries in cosmology and molecular biology. He was born in Odessa, the son of a schoolteacher, and educated at Leningrad University. He obtained his doctorate in 1928 and became professor of physics in 1931. After visiting Germany, Denmark, and England, he emigrated to the United States in 1934 and became professor of physics at George Washington University (1934–55) and then at the University of Colorado (1956–68). His first major original contribution came in 1928, when he used quantum mechanics to explain how alpha particles are emitted in radioactive decay.

In cosmology Gamow developed the 1927 theory of Belgian astronomer **Georges Lemaître** (1894–1966) about the big bang origin of the Universe. In 1948, together with American physicist Ralph Alpher (1921–) and German-born Hans Bethe (1906–2005), Gamow announced the famous alpha, beta, gamma (Alpher, Bethe, Gamow) theory. It postulates that the Universe began in a gigantic explosion of a primeval "atom" and that the resulting fireball is still expanding. Gamow estimated that the explosion occurred about 17 billion years ago (today's estimates range from 10 to 15 billion years). He also predicted the existence of cosmic background radiation, confirmed in 1964 by American radio astronomers Robert Wilson (1936–) and Arno Penzias (1933–).

In the 1950s Gamow turned his attention to the newly discovered DNA (deoxyribonucleic acid) molecule, which consists of a chain of nucleic acid bases. He postulated that the varying order of three of the four bases represents a "code" for arranging the 20 or so amino acids to form proteins in cells. The genetic code has become one of the established principles of molecular biology. Gamow also wrote a number of books in which he explained particle physics, quantum theory, and relativity to a nonscientific audience.

Goddard, Robert (1882–1945)

Robert Hutchings Goddard was an American inventor and physicist who became the country's first rocket engineer, although his achievements were not properly recognized during his lifetime. He was born in Worcester, Massachusetts, and educated at Worcester Polytechnic Institute. He graduated in 1908 and earned a doctorate in physics from Clark University in 1911, where he taught physics for many years.

Goddard experimented with rocket propellants and by 1924 tested a liquid-fuel rocket motor, using liquid oxygen and gasoline. In 1926 he successfully launched the world's first liquid-fuel rocket from his aunt's farm in Auburn, Massachusetts. With a grant from the Smithsonian Institution in Washington, D.C. he built a rocket for carrying instruments to high altitudes (launched in 1929). Financial help from the Guggenheim Foundation in 1930 enabled him to launch rockets that reached a height of about 2,000 feet (600 m) and a speed of 500 miles per hour (800 km/h). By then he had moved to a ranch near Roswell, New Mexico. His rockets of 1935 traveled faster than the speed of sound, and two years later he launched rockets that reached an altitude of nearly 10,000 feet (3,000 m).

In 1942 Goddard worked for the U. S. Navy at Annapolis, Maryland, designing rockets to provide rocket-assisted takeoff for airplanes on carriers. Despite Goddard's war effort, the American government seemed unimpressed by his work, although it inspired the German rocketeers who were later welcomed so warmly into the United States after World War II. His contribution was finally recognized when the National Aeronautics and Space Administration (NASA) named the Goddard Spaceflight Center near Washington, D.C., for him.

Landsteiner, Karl (1868–1943)

Karl Landsteiner was an Austrian-born American pathologist who discovered the major human blood groups. He was born in Vienna, where he studied chemistry and qualified as a doctor in 1891. In Vienna he was employed as a pathologist at the university, becoming professor of pathology in 1911. In 1919 he went to work in The Hague, Netherlands, and after two years he went to the Rockefeller Institute in New York, where he worked until he died.

While working as an assistant in Vienna, Landsteiner became interested in serology (the study of blood serum) and questioned why blood transfusions so often failed. In 1900 he published a paper showing that people can be split into three groups with blood types A, B, and C (the C was later changed to O). People with type A blood can receive a successful transfusion from a person with type A (but not type B), while people with type B blood need type B (not type A). Anyone can receive type O blood, known as the universal donor. In 1902 Landsteiner's coworkers found a fourth blood group, type AB, which is the universal recipient (meaning that people with type AB blood can receive blood from all the other types). In 1927 Landsteiner discovered the factors M and N in human blood. He also did other important work. In 1908 he showed that polio is caused by a virus, and in 1940 (by now aged 72) he identified the rhesus (Rh) factor in blood.

Lemaître, Georges (1894–1966)

Georges Lemaître was a Belgian astronomer and cosmologist who inspired a theory for the origin of the Universe, later known as the big bang theory. He was born in southwest Belgium and studied engineering at Louvain University. He served in the army during World War I and then took up the study of mathematics and physical sciences, receiving his doctorate in 1920. He was ordained as a Catholic priest in 1923 after attending a seminary and was awarded a traveling scholarship that took him to Cambridge (England), Harvard, and the Massachusetts Institute of Technology.

Lemaître became acquainted with the observation by American astronomer Vesto Slipher (1875–1969) of redshifts in the spectra of galaxies (implying that they are receding from Earth), and in 1927 he published a paper proposing a theory of an expanding Universe. In the same year he became professor of astronomy at Louvain, where he remained until his retirement. From 1945 onward he advocated the idea of an unstable "primeval atom" that exploded (the big bang) and started the expanding Universe. Lemaître called it his "fireworks theory."

Lorenz, Konrad (1903–89)

Konrad Zacharias Lorenz was an Austrian zoologist who founded the science of ethology, which emphasizes the study of animal behavior in an animal's natural habitat. He was born in Vienna, the son of an orthopedic surgeon, and studied medicine at Columbia University in New York and also in Vienna, qualifying in 1929. He chose a zoological subject for his doctorate of 1933 and in 1937 became lecturer in animal psychology at Vienna University. He served as a physician in the German army during World War II and was captured by the Russians in 1944. He was released in 1948, and in 1950 he established a department of comparative ethology at Buldern in northwestern Germany for the **Max Planck** Institute. He retired in 1973 and established his own research institute in Grunau, Austria.

Lorenz began studying animal behavior (of jackdaws) in the early 1930s on his father's estate near Vienna and reared geese in order to study the chicks. Like other chicks, goslings "imprint" on their mother, who is usually the first moving object they see after hatching. (Imprinting is the process by which a newborn animal receives visual and aural stimuli from its parent, which will affect its subsequent adult behavior.) Lorenz allowed the birds to imprint on him, and he could often be seen with a train of chicks following him around. By 1937 Lorenz proposed that much complex instinctive behavior in animals comes "ready formed" and does not have to be learned. He also described releaser mechanisms, in which a specific stimulus brings out an inborn behavior (for example, the red breast of a robin releasing aggression in rival males).

Messerschmitt, Willy (1898–1978)

Wilhelm E. Messerschmitt was a German aircraft engineer and designer whose company built a series of successful fighter planes for the Luftwaffe during World War II. He graduated from the Munich Institute of Technology in 1923. Within three years he was working as chief designer at Bayerische Flugzeugwerke (a Bavarian airplane factory). His early designs concentrated on gliders and light airplanes because there was a ban on the production of military aircraft in Germany in the years after World War I.

In 1932 Messerschmitt bought out his employer, who was bankrupt, and two years later he designed a small civil airplane, the Bf-108 *Taifun* (Typhoon). Then in 1935 he produced the Bf-109 fighter, which saw service during the Spanish Civil War (1936–39). With the company's name change to Messerschmitt Aktien-Gesellschaft the plane became the Me-109 and earned the world speed record in 1939 of 469 miles per hour (755 km/h). By the end of World War II German factories had produced nearly 35,000 Me-109s. The company also produced the twin-engined Me-110 night fighter/bomber and the twin-jet Me-262, the first jet plane to go into service. Less successful was the 1944 Me-163B Komet, a rocket-propelled piloted interceptor that could fly at 550 miles per hour (880 km/h) for just 10 minutes.

The Americans detained Messerschmitt for two years at the end of the war, and then in 1952 he became adviser on aviation to the Spanish government. From 1953 to 1962 the German Messerschmitt factory made three-wheeler bubble cars.

Millikan, Robert (1868–1953)

Robert Andrews Millikan was an American physicist who first measured the electric charge on an electron. He was born in Morrison, Illinois, son of a Congregational minister. He studied classics at Oberlin and went to Columbia University, obtaining his doctorate in 1895. He studied in Europe for a year with German physicists **Max Planck** (1858–1947) in Berlin and Walther Nernst (1864–1941) in Göttingen before going to the University of Chicago as assistant to German-born American physicist Albert Michelson (1852–1931) in 1896, becoming professor there in 1910. In 1921 he moved to the California Institute of Technology as director of the Norman Bridge Laboratory, where he remained until he retired in 1945.

Millikan began his work on the electron at Chicago in 1909. He gave two horizontal metal plates opposite electric charges (creating an electric field between them) and sprayed fine oil droplets between the plates. From their rate of fall under the force of gravity he calculated the masses of the oil drops. He then used X-rays to ionize the air and give electric charges to the droplets. The charged plates attracted the droplets of opposite charge, and from the changes in the rates of fall and the strength of the electric field he calculated the charge on the droplets. He found that each drop had a charge that was a simple multiple of one basic charge, which must be the charge on an electron. The value, 4.774×10^{-10} electrostatic units, remained the accepted value for 20 years. From 1923 Millikan studied cosmic rays, showing that they originate in outer space. He was wrong, however, when he theorized that the rays were electromagnetic radiation (like X-rays and light). Their true identity as particles was later proven by American physicist **Arthur Compton** (1892–1962) in 1938.

Pauli, Wolfgang (1900–58)

Wolfgang Pauli was an Austrian-born Swiss theoretical physicist who formulated the Pauli exclusion principle concerning the permitted quantum states of electrons in atoms. He was born in Vienna, son of a professor of physical chemistry, and he studied at Munich University, gaining his doctorate when he was only 21 years old. He worked with German physicist Max Born (1882–1970) at Göttingen University and with Danish physicist **Niels Bohr** (1885–1962) in Copenhagen before accepting a professorship at Hamburg

University in 1923. Five years later he moved to Zurich and became professor of physics at the Federal Institute of Technology, where he remained until his death, although he spent the years of World War II (1939–45) at Princeton.

In about 1924 Pauli added a fourth "spin" quantum number to the existing three to account for the anomalous Zeeman effect (the splitting of a spectral line into two or more components of slightly different frequency when the light source is placed in a magnetic field) and in 1925 revealed his exclusion principle. It states that no atom can have two electrons with the same quantum numbers. For example, if one electron has a spin of $+\frac{1}{2}$, the other must have a spin of $-\frac{1}{2}$. In 1930 he postulated the existence of the neutrino, a neutral low-mass subatomic particle that must, he said, be produced during beta decay, the radioactive decay in which a neutron changes into a proton and a beta particle (electron).

Perey, Marguerite (1909–75)

Marguerite Catherine Perey was a French chemist who discovered the radioactive element francium. She studied at the Paris Faculté des Sciences. In 1929 she went to work for French physical chemist Marie Curie (1867–1934) at the Radium Institute. She moved to the University of Strasbourg in 1940, became professor of nuclear chemistry in 1949, and was appointed director of the Center for Nuclear Research in 1958 .

The element with atomic number (neutron number) 87 represented one of the unfilled gaps in the periodic table in 1939 when Perey began studying the radioactive decay of actinium-227. She found that it emitted alpha particles (helium nulcei) to form an element of mass 223. She called the new element actinium K but in 1945 changed the name to francium in honor of her native land.

Schrödinger, Erwin (1887–1961)

Erwin Schrödinger was an Austrian theoretical physicist, famous for formulating the wave equation of the hydrogen atom—a complete mathematical description of the atom in terms of its mass and energy. He was born in Vienna, son of a factory owner, and obtained his doctorate from Vienna University in 1910. He worked as a researcher at the university until the outbreak of World War I, when he enrolled in the Austrian army as an artillery officer. He held various professorships in Europe before going to Berlin University in 1927, which he left in 1933 to avoid the Nazis. He went to England but returned to Austria three years later, only to leave again after Germany invaded Austria in 1938 to go to the Institute of Advanced Studies in Dublin, Ireland.

The starting point for Schrödinger's theories was the wave-particle duality proposed in 1924 by French physicist **Louis de Broglie** (1892–1987), who postulated that certain entities, such as electrons, can sometimes behave like waves and sometimes like particles. Schrödinger's wave equation of 1926 describes such systems mathematically in the branch of theoretical chemistry that came to be called wave mechanics.

Tombaugh, Clyde (1906–97)

Clyde William Tombaugh was an American astronomer who discovered the planet Pluto. He was born in Streator, Illinois, the son of a farmer. Without a college education he built his own telescope and taught himself enough to get a job in 1929 at the Lowell Observatory in Flagstaff, Arizona. He won a scholarship to the University of Kansas in 1933, graduating in 1936. In 1946 he moved to the Aberdeen Ballistic Laboratories in New Mexico, becoming an astronomer in 1955 and a professor at New Mexico University ten years later. He retired in 1973.

The existence of a planet beyond the orbits of Uranus and Neptune, named Planet X, had been predicted in 1905 by American astronomer Percival Lowell (1855–1916) to account for irregularities in Neptune's orbit. Beginning in 1929, Tombaugh studied hundreds of sky photographs using a blink comparator, which detects any object that has moved between the taking of two photographs a few days apart. He observed Pluto in early 1930 as an extremely faint object in the constellation Gemini. He failed to find any other trans-Neptunian objects despite looking for another 15 years and examining 90 million images of stars. However, he did find a comet, a nova, several asteroids, and numerous new star clusters. After World War II he constructed telescopic cameras for tracking rockets launched from the New Mexico Proving Ground.

Turing, Alan (1912–54)

Alan Mathison Turing was an English mathematician and logician who put forward original theories about computers and their processes. He was born in London, went to school in the southwest of England, and then to Cambridge University. As a postgraduate in 1935 he turned to mathematical logic, and a year later he devised the "Turing machine"—a universal hypothetical computer that never fails. It has infinite memory capacity and is therefore the basis of the modern digital computer. A Turing machine has a tape divided into segments. Each segment contains a symbol, either "0" or "1." The machine has a read-write head that can move left and right along the tape to scan successive segments. When it reaches a passive state and stops working, its output is what is left on the tape.

After a short period at Princeton University Turing returned to England. In 1939 he joined the team that cracked the German military code named Enigma. In 1945 he helped develop ACE (Automatic Computing Engine) at the National Physical Laboratory and three years later went to Manchester University to do similar work on MADAM (Manchester Automatic Digital Machine), one of the world's first successful computers and the one with the largest storage capacity.

In 1950 he proposed a way of telling the difference between a computer and an intelligent human, called the Turing Test. In the test an isolated interrogator asks questions either of a computer (in another room) or a human (in a third room); the interrogator does not know which is which. If the interrogator cannot tell the difference, the computer has reached the ability to think like a human. This was the original research into artificial intelligence.

Turing began to study morphogenesis—the way that patterns form and change in the natural world—in 1952. The work was interrupted when he was prosecuted for homosexuality and sentenced to compulsory drug treatment, which led to depression. He took his own life in 1954 by cyanide poisoning.

Urey, Harold (1893–1981)

Harold Clayton Urey was an American chemist whose main achievement was the discovery of deuterium (heavy hydrogen), an isotope of hydrogen of mass 2 ("normal" hydrogen has mass 1). He was born in Walkerton, Indiana, the son of a teacher. He studied zoology at Montana State University and chemistry at the University of California at Berkeley, getting his doctorate in 1923. He worked with Danish physicist **Niels Bohr** (1885–1962) at the Institute of Theoretical Physics in Copenhagen for a year and in 1924 became associate in chemistry at Johns Hopkins University. He moved to Columbia University, New York, in 1929 and stayed there for nearly 30 years before joining the University of California at San Diego.

In 1932 Urey evaporated liquid hydrogen to separate out deuterium (which evaporates more slowly than normal hydrogen), proving its existence spectroscopically. During World War II he worked on separating uranium isotopes for the Manhattan Project to make the atom bomb. He also developed a large-scale process for making heavy water (deuterium oxide) as a moderator for nuclear reactors. After the war he turned his attention to geochemistry, using the relative amounts of oxygen isotopes in fossil seashells to determine the temperature of the seawater in which they grew.

Whittle, Frank (1907–96)

Frank Whittle was a British aeronautical engineer working as a test pilot for the Royal Air Force (RAF) when he patented the first turbo jet engine in 1930. Nobody could guess at that time the significance the lightweight jet engine would have on the future development of airplanes, helicopters, and even missiles. Whittle was born in the central England city of Coventry, the son of a mechanic. He joined the RAF as an apprentice in 1923, training at the RAF College, Hanwell, and at Cambridge University, where from 1934 to 1937 he studied mechanical sciences.

A jet engine basically consists of a pear-shaped tube with a rotating shaft along its length. A rapidly turning wheel set with vanes, like a turbine, compresses air as it enters the front of the engine and forces it into the combustion chamber. There the air/fuel mixture—the usual fuel is kerosene—burns to produce a large volume of hot gases. As the gases leave the rear of the engine, they turn a second turbine mounted on the same shaft, and in this way they also turn the compressor. The principle is simple, but it took somebody of Whittle's ability to think it through and then build an engine that worked.

The high temperatures involved made it difficult to find alloys that were tough enough to construct the engine. Not until after 1936, when Whittle and his associates formed the Power Jets company, was real progress made. The first Whittle engine was tested in 1937, and the first British airplane equipped with a turbo jet, the Gloster Whittle E-28, flew in May 1941; the small engine had a thrust of only 827 pounds (375 kg). The makers sent a prototype engine to the United States, where it became the basis for the engine in the experimental Bell XP-59A jet plane.

By the late stages of World War II in 1945 jet planes were just coming into service with the RAF (which had the twin-engined Gloster Meteor) and the U. S. Air Force. Whittle became a Fellow of the Royal Society in 1947 and received a knighthood after he retired from the RAF in 1948.

Unknown to Whittle, in 1936 a 25-year-old German engineer named Hans von Ohain (1911–98) conducted similar experiments on behalf of his employer, the airplane manufacturer Heinkel. He had taken out a patent in 1935. The first Heinkel He-178 jet plane flew in August 1936. Ohain went on in 1942 to design the twin-engined Messerschmitt-262 jet fighter for the Luftwaffe, but it did not go into large-scale production because by that time the German authorities had decided to divert much of their war effort to rockets.

Zworykin, Vladimir (1889–1982)

Vladimir Kosma Zworykin was a Russian-born American physicist and electronics engineer who is best remembered for his invention of the television camera tube called the iconoscope. He also made a major contribution to the development of the electron microscope. He was born in Murum near Moscow, the son of a riverboat merchant. He studied electrical engineering at the Petrograd (now St. Petersburg) Institute of Technology and graduated in 1912. After working in Paris under French physicist Paul Langevin (1872–1946), he joined the Russian army as a radio officer during World War I. In 1919 he moved to the United States and a year later joined the Westinghouse Electric Corporation in Pittsburgh and obtained a doctorate from the university there in 1926. He became a U. S. citizen in 1924. From 1929 he worked for RCA (Radio Corporation of America), where in 1946 he became director of research and eventually vice president until 1954.

Zworykin invented the iconoscope, which used electronic scanning, in 1923 (which was eventually patented in 1938). It employed magnets to make the electron beam scan from side to side. From this it was a logical extension of the idea to make magnets focus an electron beam in much the same way as a lens focuses light. In their wave mode—electrons can be thought of either as streams of particles or as a wave—electrons behave like light of extremely short wavelength.

By 1939 Zworykin had embodied these ideas in the first practical electron microscope. Previously, the wavelength of light limited the resolving power of a microscope—Abbe's law states that an optical microscope cannot distinguish between two objects if their distance apart is less than half the wavelength of the light illuminating them. But with a beam of high-energy electrons for illumination objects as small as individual molecules can now be "seen" (or at least recorded on a phosphorescent screen, photographic plate, or computer display). During World War II (after 1941) Zworykin's work at RCA concerned aircraft fire control, radar, and television-guided missiles.

GLOSSARY

Any of the words printed in SMALL CAPITAL LETTERS can be looked up in this Glossary.

Abbreviations
b. Born (of a deceased person for whom only the birth date is known).
c. About (from Latin *circa*).
d. Died (of a deceased person for whom only the death date is known).
date with dash (–), e.g., Bill Gates 1955– ,used to signify that, at the time of writing, the person was still living.
fl. Flourished (from Latin *floreat*, of a person whose birth and death dates are not known).

alpha particle A positively charged, high-energy particle emitted from the NUCLEUS of a radioactive atom; a helium nucleus.

amplifier An electronic device that increases the strength of an input current or voltage.

antineutrino The ANTIPARTICLE of a NEUTRINO.

antiparticle A particle that has properties opposite to those of the corresponding ordinary particle. For example, an antineutrino is like a NEUTRINO, but negatively charged.

big bang theory Widely accepted theory proposed by Belgian astronomer George Lemaître (1894–1966) that the Universe began as a "primal atom," that a massive explosion caused it to expand outward, and that it is still expanding.

cathode The negative electrode of a battery or similar device, through which an electric current passes.

cathode ray A stream of ELECTRONS emitted by a CATHODE when heated.

centrifuge An apparatus that spins around to separate particles from a suspension.

chain reaction A series of nuclear reactions in heavy atoms (such as uranium or plutonium) in which a NUCLEUS splits, releasing several NEUTRONS, which collide with other nuclei and cause them to split, and so on.

corona The outer part of the Sun's atmosphere.

cosmic ray A subatomic particle, usually a PROTON, moving through space at close to the speed of light.

cyclotron A device that accelerates atomic particles to high speeds by making them follow a spiral path between the poles of two D-shaped magnets.

deuteron A SUBATOMIC PARTICLE; the NUCLEUS of a hydrogen-2 (deuterium) atom consisting of one NEUTRON and one PROTON.

electron A negatively charged SUBATOMIC PARTICLE.

electron microscope A type of microscope that uses a stream of ELECTRONS and electromagnets to produce highly magnified images.

electrophoresis A method of separating charged particles in a fluid or gel by passing an electric current through electrodes dipping into the fluid.

elementary particle Any one of five stable particles that make up atoms, molecules, and all matter in the Universe.

enzyme A large PROTEIN molecule that acts as a catalyst for the chemical reactions on which life depends.

fissile Capable of undergoing NUCLEAR FISSION.

game theory A branch of applied mathematics involving the study of decision-making in situations where strategic interaction (moves and countermoves) occurs between opponents.

gamma rays High-energy PHOTONS emitted from ATOMIC NUCLEI during radioactive decay.

hemoglobin The oxygen-carrying molecule in vertebrate red blood cells.

holography An advanced form of photography in which an image is recorded in three dimensions.

hormone A chemical substance secreted in one part of an organism and transported to another part, where it produces a response.

ionization A process during which ELECTRONS gain sufficient energy to escape completely from an atom or molecule.

isotope Atoms of the same element that have the same atomic number but different numbers of NEUTRONS, thus differing in their relative atomic masses.

lepton An elementary particle, such as an ELECTRON or NEUTRINO, that does not take part in interactions involving the STRONG NUCLEAR FORCE and is not composed of QUARKS.

lobotomy The practice (now generally considered inappropriate) of performing surgery on part of the brain to treat pain or mental disorders.

meson A particle composed of a QUARK and an antiquark. It can have a positive, negative, or zero charge.

muon A LEPTON with a negative charge.

neutrino An ELEMENTARY PARTICLE that possesses a tiny mass but no charge.

neutron One of the three main SUBATOMIC PARTICLES. They carry no electric charge and they occur in the NUCLEI of all atoms except hydrogen.

nuclear fission A process by which an ATOMIC NUCLEUS splits into lighter elements.

nuclear fusion A process by which two or more ATOMIC NUCLEI join together to make a heavier one.

nuclear reactor A device for generating energy either by NUCLEAR FISSION or by NUCLEAR FUSION.

nucleon Either of the two components of the NUCLEUS of an atom (a PROTON or a NEUTRON).

nucleus, atomic (pl. nuclei) The positively charged dense region at the center of an ATOM, composed of PROTONS and NEUTRONS.

photon The elementary particle of energy in which light and other forms of electromagnetic radiation are emitted.

polymer Any compound of large molecules formed by many smaller molecules (monomers) combining in a regular pattern.

positron The antiparticle (counterpart) of an ELECTRON, having the same mass but positive charge.

proton A positively charged SUBATOMIC PARTICLE in the ATOMIC NUCLEUS.

quantum electrodynamics The branch of physics that studies the properties of electromagnetic radiation and the ways in which it interacts with charged particles.

quantum number One of a set of four numbers that uniquely characterize an ELECTRON and its state in an atom. They relate to energy levels, momentum, and spin.

quantum theory Theory developed in the early 20th century based on the suggestion by German physicist Max Planck (1858–1947) that light is emitted in separate packets, or quanta, also known as PHOTONS.

quark The elementary SUBATOMIC PARTICLE that is the fundamental component of NEUTRONS and PROTONS, but not ELECTRONS. For each quark there is an antiparticle (counterpart), called an antiquark.

radioactivity The spontaneous disintegration of certain unstable NUCLEI, accompanied by the emission of ALPHA PARTICLES, beta rays (ELECTRONS), or GAMMA RAYS.

redshift The lengthening of the wavelength of SPECTRAL LINES, caused either by the motion of the source away from the observer or by the motion of the observer away from the source.

semiconductor A material that has a resistance intermediate between those of an insulator and a conductor.

solid-state electronics The science and technology of using SEMICONDUCTOR materials rather than VACUUM TUBES to make electronic circuits and devices.

spectral lines The dark lines in a continuous SPECTRUM that are produced when electromagnetic radiation passes through a gas cloud, and certain wavelengths are absorbed by the ELECTRONS around the gas atoms.

strong nuclear force The force of nature through which SUBATOMIC PARTICLES, such as MESONS and PROTONS, communicate with each other. It is the strongest of all forces but acts only over the distance of an ATOMIC NUCLEUS.

subatomic particle Any particle that is smaller than an atom.

superconductor A substance that shows no resistance to the passage of electric current (usually at very low temperatures).

supergiant A star that typically has 100 times the luminosity and a larger radius than a giant star of the same spectral classification.

supernova A catastrophic explosion that blows a star to pieces.

tracer A compound in which a stable atom is replaced by a radioactive ISOTOPE of the same ELEMENT to make its path through a biological system traceable.

transistor A SOLID-STATE electronic device that amplifies a small signal current or voltage and turns it into a large output current or voltage.

ultracentrifuge A high-speed CENTRIFUGE used to determine the relative molecular masses of large molecules in high POLYMERS and proteins.

vacuum tube An airtight glass tube in which electricity is conducted by ELECTRONS passing through a partial vacuum from a CATHODE to an anode; also called an electron tube.

vitamin A nutrient substance that is essential, in small quantities, for health.

white dwarf A type of small, dense star of relatively low temperature and brightness. "Normal" stars end their lives as white dwarfs.

X-ray Penetrating electromagnetic radiation of very short wavelength and high energy; widely used in medicine for the diagnosis and treatment of disorders.

SET INDEX

A **bold** number shows the volume and is followed by the relevant page numbers (e.g., 1:11, 36). Page numbers underlined (e.g. 7:24-5 indicate the main entries for that topic. Italic page numbers (e.g., 2:*21*) refer to illustrations and their captions. Page numbers followed by an asterisk (e.g. 3:50*) refer to entries listed in the Key People pages.

D

M

N

Further Reading

General

Allaby, M., and D. Gjersten, *Makers of Science*, Oxford University Press, 2002.

Asimov, A., *Asimov's Biographical Encyclopedia of Science and Technology*, Avon Books, 1972.

Boorstin, D. J., *The Discoverers*, Random House, 1983.

Boyles, D., *The Tyranny of Numbers*, HarperCollins, 2000.

Bunch, B., and A. Hellemans, *The Timetables of Science*, Simon and Schuster, 1988.

Bunch, B., and A. Hellemans, *The Timetables of Technology*, Simon and Schuster, 1993.

Carey, J. (ed.), *The Faber Book of Science*, Faber and Faber, 1995.

Crystal, D. (ed.), *The Cambridge Biographical Dictionary*, Cambridge University Press, 2000.

Daintith, J. (ed.), *A Dictionary of Scientists*, Oxford University Press, 1999.

Day, L., and I. McNeil (eds.), *Biographical Dictionary of the History of Technology*, Routledge, 1998.

Diamond, J., *Guns, Germs and Steel*, Vintage, 1998.

The Dictionary of National Biography, Oxford University Press, 1982.

Dyson, J., and R. Uhlig (ed.), *A History of Great Inventions*, Robinson, 2002.

le Fanu, J., *The Rise and Fall of Modern Medicine*, Little, Brown and Company, 1999.

Giscard d'Estaing, V-A., *The Book of Inventions and Discoveries*, Macdonald Queen Anne Press, 1990.

Gribbin, J., *Science, A History*, BCA, 2002.

Harrison, I., *The Book of Inventions*, Cassell, 2004.

Hoskin, M. (ed.), *The Cambridge Concise History of Astronomy*, Cambridge University Press, 1999.

The Hutchinson Dictionary of Scientific Biography, Helicon, 2000.

The Inventions that Changed the World, Reader's Digest, 1982.

Margotta, R., *The History of Medicine*, Octopus, 1996.

Meadows, J., *The Great Scientists*, Oxford University Press, 1997.

Messadié, G., *Great Inventions through History*, Chambers, 1991.

Messadié, G., *Great Scientific Discoveries*, Chambers, 1991.

Millar, D., et al., *The Cambridge Dictionary of Scientists*, Cambridge University Press, 1996.

Muir, H. (ed.), *Larousse Dictionary of Scientists*, Larousse, 1994.

Parry, M. (ed.), *Chambers Biographical Dictionary*, Chambers, 1997.

Philip's Astronomy Encyclopedia, George Philip Limited, 2002.

Philip's Science & Technology Encyclopedia, George Philip Limited, 1998.

Philip's Science & Technology People, Dates & Events, George Philip Limited, 1999.

Porter, R., *The Greatest Benefit to Mankind*, HarperCollins, 1997.

Silver, B. L., *The Ascent of Science*, Oxford University Press, 1998.

Tallack, P (ed.), *The Science Book*, Cassell, 2001.

Trefil, J., *Cassell's Laws of Nature*, Cassell, 2002.

Waller, J., *Fabulous Science*, Oxford University Press, 2002.

Webster's Biographical Dictionary, G. & C. Merriam, 1971.

What Happened When?, HarperCollins, 1994.

Whitfield, P., *Landmarks in Western Science*, The British Library, 1999.

Williams, T. I. (ed.), *A Biographical Dictionary of Scientists*, Adam & Charles Black, 1974.

Williams, T. I., *A History of Invention*, Time Warner Books, 2003.

Specific to this Volume

Brockman, J. (ed.), *The Greatest Inventions of the Past 2,000 Years*, Weidenfeld & Nicolson, 2000.

Burderi, R., *The Invention that Changed the World*, Simon & Schuster, 1996.

Emsley, J., *Nature's Building Blocks*, Oxford University Press, 2001.

Fisher, D. E., and M. J. Fisher, *The Invention of Television*, Counterpoint, 1996.

Macfarlane, G., *Alexander Fleming: The Man and the Myth*, Chatto & Windus, 1984.

Matthews, R., *The Illustrated History of the 20th Century*, Grenville Books Ltd, 1992.

Messadié, G., *Great Modern Inventions*, Chambers, 1991.

Thompson, L., *Guns in Colour*, Octopus, 1981.

Who's Who in the Twentieth Century, Oxford University Press, 1999.

Woodman, R., *The History of the Ship*, Conway Maritime Press, 1997.

Terminology Reference

Clark, J. O. E., and S. Stiegler (eds.), *The Facts on File Dictionary of Earth Science*, Checkmark Books, 2000.

Daintith, J. (ed.), *A Dictionary of Chemistry*, Oxford University Press, 2000.

Daintith J., and J. Clark (eds.), *The Facts on File Dictionary of Mathematics*, Facts on File, 1999.

Darton, M., and J. Clark, *The Macmillan Dictionary of Measurement*, Macmillan, 1994.

Illingworth, V. (ed.), *Collins Dictionary of Astronomy*, HarperCollins, 1994.

Waites, G., *The Cassell Dictionary of Biology*, Cassell, 1999.

Useful Web Sites

http://www.atomicmuseum.com/
Web site of the National Atomic Museum (soon to be renamed the National Museum of Nuclear Science and History), Albuquerque, New Mexico.

http://www.howstuffworks.com
A comprehensive Web site that gives detailed explanations of how everything around us actually works.

http://www.mhs.ox.ac.uk/
Web site of Oxford University's Museum of the History of Science.

http://www.nasa.gov/

Official Web site of the National Aeronautics and Space Administration (NASA) with links to learning resources and information for students, and full details of all missions.

http://www.nasm.si.edu/
Web site of the Smithsonian National Air and Space Museum, Washington, D.C.

http://www.nscdiscovery.org/
Web site of the National Science Center, Augusta, Georgia.

http://nobelprize.org/
Official Web site of the internatonal Nobel Prize awards given annually since 1901 for achievements in physics, chemistry, medicine, literature, and for peace. Click on "list of all prizewinners" and then on individual links to read about award-winning achievements.

http://www.psigate.ac.uk/
Physical Sciences Information Gateway provides access to Web resources in the physical sciences, including astronomy, chemistry, earth sciences, materials sciences, physics, and general science.

http://whyfiles.org/
University of Wisconsin Web site that explains the science behind the news.